T0205477

T-Labs Series in Telecommunication Services

Series Editors

Sebastian Möller, Quality and Usability Lab, Technische Universität Berlin, Berlin, Germany

Axel Küpper, Telekom Innovation Laboratories, Technische Universität Berlin, Berlin, Germany

Alexander Raake, Audiovisual Technology Group, Technische Universität Ilmenau, Ilmenau, Germany

It is the aim of the Springer Series in Telecommunication Services to foster an interdisciplinary exchange of knowledge addressing all topics which are essential for developing high-quality and highly usable telecommunication services. This includes basic concepts of underlying technologies, distribution networks, architectures and platforms for service design, deployment and adaptation, as well as the users' perception of telecommunication services. By taking a vertical perspective over all these steps, we aim to provide the scientific bases for the development and continuous evaluation of innovative services which provide a better value for their users. In fact, the human-centric design of high-quality telecommunication services – the so called "quality engineering" – forms an essential topic of this series, as it will ultimately lead to better user experience and acceptance. The series is directed towards both scientists and practitioners from all related disciplines and industries.

** Indexing: books in this series are indexing in Scopus **

More information about this series at http://www.springer.com/series/10013

Felix Beierle

Integrating Psychoinformatics with Ubiquitous Social Networking

Advanced Mobile-Sensing Concepts and Applications

 Springer

Felix Beierle (iD)
Technische Universität Berlin
Berlin, Germany

Zugl.: Berlin, Technische Universität, Diss., 2020 u. d. T. *Integrating Psychoin-formatics with Ubiquitous Social Networking Through Advanced Mobile-Sensing Concepts and Applications*

ISSN 2192-2810   ISSN 2192-2829   (electronic)
T-Labs Series in Telecommunication Services
ISBN 978-3-030-68842-4   ISBN 978-3-030-68840-0   (eBook)
https://doi.org/10.1007/978-3-030-68840-0

This Springer imprint is published by the registered company Springer Nature Switzerland AG
The registered company address is: Gewerbestrasse 11, 6330 Cham, Switzerland

Integrating Psychoinformatics with Ubiquitous Social Networking

Advanced Mobile-Sensing Concepts and Applications

vorgelegt von
M.A., M.Sc.
Felix Beierle
ORCID: 0000-0003-2702-9893

an der Fakultät IV – Elektrotechnik und Informatik
der Technischen Universität Berlin
zur Erlangung des akademischen Grades

Doktor der Naturwissenschaften
- Dr. rer. nat. -

genehmigte Dissertation

Promotionsausschuss:

Vorsitzender: Prof. Dr.-Ing. Sebastian Möller, Technische Universität Berlin

Gutachter: Prof. Dr. Axel Küpper, Technische Universität Berlin

Gutachter: Prof. Dr. Rüdiger Pryss, Julius-Maximilians-Universität Würzburg

Gutachter: Prof. Dr.-Ing. Jöran Beel, Trinity College Dublin

Tag der wissenschaftlichen Aussprache: 23. Juni 2020

Berlin 2020

Abstract

There are fundamental differences in the way people experience behaviors and emotions. The field of *psychometrics* allows the quantifiable measurement of *personality traits* that describe such inter-personal differences. Vast amounts of psychological and psychoinformatical studies show that personality traits are not only associated with different everyday preferences, but also with the way people form interpersonal bonds in social networks. However, these research results are rarely considered in the research and development of social networking services. We argue that such understanding of people, their behaviors, emotions, and how they form bonds with other people can be integrated in social networking services and may ultimately improve real-life social well-being. The objective of this doctoral thesis thus is the integration of *psychoinformatics*—the intersection of psychology and computer science—with ubiquitous social networking.

Psychoinformatics uses objective measurements to research and deepen the understanding of personality-related differences in human experiences and behaviors. A common approach is *mobile sensing*. Such tracking of sensor data has serious implications regarding the privacy of users. However, currently, there are no structured approaches in dealing with these privacy concerns. Furthermore, when planning a psychoinformatical study, it is unclear to what extent demographical properties and personality traits play a role when collecting data from daily life scenarios. We developed a structured approach for privacy-awareness for mobile sensing in psychoinformatics and conduct a study on what data researchers can expect users to be willing to share. Our results can help researchers with study planning, and study participants benefit by increased privacy-awareness. While there have been several studies related to personality traits and smartphone data, there are knowledge gaps, specifically with regard to smartphone usage frequency and duration in relation to the user's personality. We collected and analyzed data and fill this gap. Our results are valuable for psychologists, for example, when researching smartphone overuse. Additionally, our results helped researchers and software developers of mobile systems understand their user base better.

In particular, we designed, developed, and deployed the Android app TYDR— Track Your Daily Routine—for the collection of smartphone sensor and usage data. We developed and implemented PM-MoDaC, a full-scale privacy model for mobile data collection apps, consisting of nine different technical and design-related measures. We released TYDR on Google Play and attracted nearly 4,000 users. Evaluating our privacy model, we analyzed what data users are willing to share with researchers; for example, younger users tend to be less willing to share data. We conducted a study analyzing the relationship between personality traits and smartphone usage frequency and usage session duration. On average, the users in our sample (n = 526) used their smartphone 72 times per day, with a mean session duration of 3.7 minutes. Our study reveals that neurotic and extraverted users use their phones more frequently, while conscientious users have shorter session durations.

We integrated psychoinformatics with ubiquitous social networking by developing concepts, metrics, and applications that are based on mobile sensing and consider psychological insights about real-life social networking. Most existing systems rely on manually entered profile data, making them tedious to use. Our results are highly automated unobtrusive ubiquitous social networking systems that may help improve social well-being by providing smartphone-mediated incentivizations of social interaction, and by seamlessly providing services to individual users and groups of users. Users benefited from our systems' high degree of automation, their unobtrusiveness, and from potentially more meaningful smartphone-mediated real-life social networking. Our concepts, metrics, and prototypical applications serve as blueprints for researchers and developers of similar systems.

In particular, we developed SimCon, a concept for the recommendation of new contacts. Similar people in proximity are recommended based on similar smartphone data. To estimate the similarity between two users, we introduce the metric CBF-Dice utilizing probabilistic data structures. CBF-Dice is able to accurately assess similarity with a single exchange of a small Counting Bloom Filter. Furthermore, we developed two full ubiquitous social networking app prototypes. MobRec, our platform for decentralized recommender systems, is based on device-to-device data exchange and runs on both Android and iOS on off-the-shelf smartphones. Implicit preferences and explicit ratings are exchanged when users are in proximity. Local recommender systems or third-party service providers then recommend new items based on on-device data. GroupMusic, our group recommender system, is based on mobile sensing and privacy-aware sharing of data. It implements a ubiquitous computing vision: fully automatically, the system plays back music for a group of currently present users.

Zusammenfassung

Es gibt fundamentale Unterschiede in der Art und Weise, wie Menschen Verhalten und Emotionen wahrnehmen. Das Feld der *Psychometrie* erlaubt die quantifizierbare Messung von *Persönlichkeitsmerkmalen*, welche solche interpersonellen Unterschiede beschreiben. Eine erhebliche Anzahl psychologischer und psychoinformatischer Studien zeigen, dass Persönlichkeitsmerkmale nicht nur mit Alltagspräferenzen verbunden sind, sondern auch damit, wie Menschen interpersonelle Beziehungen in sozialen Netzwerken aufbauen. Jedoch werden diese Forschungsergebnisse in der Erforschung und der Entwicklung von *Social Networking* Diensten kaum in Betracht gezogen. Wir argumentieren, dass solch ein Verständnis von Menschen, von ihrem Verhalten, von ihren Emotionen und davon, wie sie Beziehungen mit anderen Menschen eingehen, in Social Networking Dienste integriert werden kann und so letztendlich das soziale Wohlergehen verbessern könnte. Das Ziel dieser Dissertation ist deswegen die Integration von *Psychoinformatik*—die Schnittmenge aus Psychologie und Informatik—und *Ubiquitous Social Networking*.

Psychoinformatik verwendet objektive Messungen, um das Verständnis von persönlichkeitsbezogenen Unterschieden in menschlichen Erfahrungen und Verhalten zu erforschen und zu vertiefen. Ein gebräuchlicher Ansatz ist *Mobile Sensing*. Solch ein Verfolgen von Sensordaten hat ernsthafte Implikationen bezüglich der Privatsphäre der Nutzer*innen, jedoch gibt es zurzeit noch keine strukturierten Ansätze, um mit diesen Privatsphärebedenken umzugehen. Des Weiteren ist bei der Planung von psychoinformatischen Studien unklar, bis zu welchem Grad demographische Eigenschaften und Persönlichkeitsmerkmale eine Rolle beim Sammeln von Daten aus dem Alltag spielen. Wir entwickeln einen strukturierten Ansatz für ein Privatsphäremodell für *Mobile Sensing* in der Psychoinformatik und führen eine Studie darüber durch, welche Bereitschaft zum Teilen von Daten Forscher*innen erwarten können. Unsere Ergebnisse helfen Forscher*innen bei der Studienplanung und Studienteilnehmer*innen profitieren von erhöhtem Datenschutzbewusstsein. Obwohl es schon einige Studien zu Smartphonedaten und Persönlichkeitsmerkmalen gibt, gibt es noch Wissenslücken, speziell in Bezug auf Smartphonenutzungsfrequenz und -nutzungssitzungsdauer im Verhältnis zur

Persönlichkeit des/der Nutzer*in. Wir sammeln und analysieren Daten und füllen diese Lücke. Unsere Ergebnisse sind wertvoll für Psycholog*innen, zum Beispiel, wenn sie übermäßige Smartphonenutzung erforschen. Zusätzlich können unsere Ergebnisse auch Forscher*innen und Softwareentwickler*innen von mobilen Systemen helfen, ihre Nutzer*innenbasis besser zu verstehen.

Wir entwerfen, entwickeln und nutzen die Android Anwendung TYDR—Track Your Daily Routine—zur Sammlung von Smartphonesensor und -nutzungsdaten. Wir entwickeln und implementieren PM-MoDaC, ein umfassendes Privatsphäremodell für Anwendungen, die sich mit der Sammlung mobiler Daten befassen. Es besteht aus neun verschiedenen technischen und entwurfsbezogenen Maßnahmen. Wir haben TYDR bei Google Play veröffentlicht und ca. 4.000 Nutzer*innen gewonnen. Wir evaluieren das Privatsphäremodell und analysieren, welche Daten Nutzer*innen bereit sind, mit Forscher*innen zu teilen. Zum Beispiel sind jüngere Nutzer*innen tendenziell weniger dazu bereit, Daten zu teilen. Wir führen eine Studie durch, die die Beziehung zwischen Persönlichkeitsmerkmalen und Smartphonenutzungsfrequenz und -dauer analysiert. Im Schnitt benutzen die Nutzer*innen unserer Stichprobe (n = 526) ihr Smartphone 72 Mal am Tag, mit einer durchschnittlichen Sitzungsdauer von 3,7 Minuten. Unsere Studie zeigt auf, dass neurotische und extravertierte Nutzer*innen ihr Smartphone öfter verwenden, während gewissenhafte Nutzer*innen kürzere Nutzungssitzungen haben.

Wir integrieren Psychoinformatik in Ubiquitous Social Networking durch das Entwickeln von Konzepten, Metriken und Anwendungen, die auf Mobile Sensing basieren und die psychologische Erkenntnisse darüber berücksichtigen, wie Menschen im echten Leben soziale Netzwerke bilden. Die meisten existierenden Systeme beruhen auf manuell eingegebenen Profildaten, was deren Nutzung mühsam macht. Unsere Ergebnisse sind höchst automatisierte, unaufdringliche, allgegenwärtige Ubiquitous Social Networking Systeme, die helfen könnten, das soziale Wohlbefinden zu verbessern, indem sie Smartphone-vermittelte Anreize für soziale Interaktionen zur Verfügung stellen und Dienste für individuelle Nutzer*innen und Gruppen von Nutzer*innen bereitstellen. Nutzer*innen profitieren von dem hohem Grad an Automatisierung unserer Systeme und ihrer Unaufdringlichkeit und von potentiell bedeutungsvolleren Smartphone-vermittelten sozialem Miteinander. Unsere Konzepte, Metriken und prototypischen Anwendungen dienen als Blaupausen für Forscher*innen und Entwickler*innen von ähnlichen Systemen.

Wir entwickeln SimCon, ein Konzept für die Empfehlung von neuen Kontakten. Menschen in der Nähe werden basierend auf der Ähnlichkeit ihrer Smartphonedaten empfohlen. Um die Ähnlichkeit von zwei Nutzer*innen abzuschätzen, führen wir die Metrik CBF-Dice ein, die probabilistische Datenstrukturen verwendet. CBF-Dice kann die Ähnlichkeit mit dem einfachen Austausch eines Counting Bloom Filters korrekt abschätzen. Des Weiteren entwickeln wir zwei vollständige Ubiquitous Social Networking Prototypen. MobRec, unsere Plattform für dezentralisierte Empfehlungssysteme, basiert auf Gerät-zu-Gerät Datenaustausch und läuft auf handelsüblichen Android- und iOS-Smartphones. Implizite Präferenzen und explizite Bewertungen werden ausgetauscht, wenn Nutzer*innen in der Nähe voneinander sind. Lokale Empfehlungssysteme oder externe Dienstanbieter*innen können dann neue Objekte auf Basis der lokalen Daten empfehlen. Unser Gruppenempfehlungssystem GroupMusic basiert auf Mobile Sensing und dem datenschutzbewussten Teilen von Daten. Es implementiert eine Vision des Ubiquitous Computing: Das System spielt vollautomatisch Musik für eine Gruppe gerade anwesender Nutzer*innen ab.

Acknowledgments

This doctoral thesis would not have been possible without other people's support, advice, and productive collaborations. First of all, I'd like to thank my supervisor Prof. Dr. Axel Küpper for his continual support and guidance over the last years. His advice and feedback was indispensable to my work. I would also like to thank Prof. Dr. Rüdiger Pryss for his helpful advice and feedback, and for always supporting me and my ideas. I also thank Prof. Dr. Jöran Beel for his valuable advice that helped me grow as a researcher, and for inviting me for a research visit to the National Institute of Informatics (NII) in Tokyo.

Working in such an interdisciplinary field as psychoinformatics, I'm thankful for the great collaborations with psychological researchers, specifically with PD Dr. Winfried Schlee, Dr. Patrick Neff, Prof. Dr. Mathias Allemand, Prof. Dr. Thomas Probst, Prof. Dr. Johannes Zimmermann, Prof. Dr. Stefan Stieger, and Dr. Sanja Budimir. I would also like to thank Prof. Dr. Yang Chen, Prof. Dr. Pan Hui, and Prof. Dr. Akiko Aizawa for their advice and for inviting me to give presentations about my work in their groups at Fudan University in Shanghai, at Hong Kong University of Science and Technology (HKUST), and at the National Institute of Informatics (NII), Tokyo.

I am grateful for all the lively discussions and mutual support with my colleagues at Technische Universität Berlin. I like to especially thank my colleagues who co-authored papers with me related to this thesis, Kai Grunert, Dr. Sebastian Göndör, and Tobias Eichinger.

Over the years, many students were part of the TYDR team. First, I'd like to thank Vinh Thuy Tran for all his efforts and his resourcefulness. I also especially thank Soumya Siladitya Mishra, Sarjo Das, Sakshi Bansal, Marcel Müller, Yong Wu, and all my other student assistants. I am also grateful for all the feedback I got from friends and colleagues during the development of TYDR. I owe a special thanks to Daniel Lenz and all TYDR users. During my time as a PhD student, I had the chance to supervise several bachelor's and master's theses. In particular, I thank Robert Staake, Jan Pokorski, Simone Egger, Viktor Schlüter, Hai Dinh Tuan, Julian Tan, and Ishan Tiwari for their great theses.

The German Federal Ministry of Education and Research (BMBF) and the Software Campus program funded my project out of which I created TYDR. The Software Campus program and its funding played a major role in my ability to pursue my own research project. I'd like to extend my thanks to Andrea Hahn and Sandra Wild for their support in navigating the intricacies of university bureaucracy.

Finally, I'd like to thank all my friends and my family, especially my parents, who always supported me in my endeavors.

Berlin, Germany   Felix Beierle

Contents

List of Figures

List of Tables

Chapter 1
Introduction

Psychological research shows that there are fundamental differences between people and how they experience behaviors and emotions. Not only does this affect individual experiences but also the forming of interpersonal relationships. Applying these findings in smartphone-mediated social networking services may help improve social well-being. This thesis aims at the integration of *psychoinformatics* and *ubiquitous social networking* and makes contributions in both of these fields.

1.1 Psychoinformatics

The emerging field of *psychoinformatics* is at the intersection of computer science and psychology. It primarily focuses on the collection and analysis of large amounts of data of objective behavioral measurements, typically from smartphone sensors or social networking sites [62, 81, 55]. Especially the smartphone enables researchers to gain a deeper understanding of detailed objective measurements in relation to psychological traits like *personality* and *mood*. Defined as a field within psychoinformatics, *digital phenotyping* extends the concept of phenotypes as observable traits to digital traces of users [45, 1]. Typical outcomes of psychoinformatical research are a deeper understanding of latent constructs like personality or mental illnesses or just-in-time interventions [1]. The technical foundation of such research is *mobile (crowd)sensing*. *Mobile sensing* is the utilization of a mobile device— typically a smartphone—and its sensors for conducting and storing measurements. *Mobile crowdsensing* refers to the measurement of large-scale phenomena that cannot be measured by single devices [34, 50, 19]. Data is collected and shared— typically with a research institute—and information is extracted. In the context of psychoinformatics and personality-related research, this means, for example, crowdsourcing data from multiple users and gaining new insights in the relationship between smartphone data and user personality.

F. Beierle, *Integrating Psychoinformatics with Ubiquitous Social Networking*,
T-Labs Series in Telecommunication Services,
https://doi.org/10.1007/978-3-030-68840-0_1

One of the key concepts in psychology in general and specifically in psychoin-formatics/digital phenotyping is *personality*. Personality relates to the behaviors, emotions, knowledge, and memory of a person [25] and shows that there are fundamental differences between people and how they experience behaviors and emotions. Vast literature of psychological research shows the connections between personality and a multitude of everyday aspects [66, 72, 48, 75, 44]. The field of *psychometrics* researches quantifiable measurements of psychological aspects. The most common tool to describe a person's personality in terms of specific traits is the Big Five model [58]. Standardized psychometric questionnaires yield concrete values for each of the traits, i.e., by filling out a questionnaire, the personality can be quantified on the five dimensions of the Big Five model. Those five dimensions are *openness to experience*, *conscientiousness*, *extraversion*, *agreeableness*, and *neuroticism*. Assessing a user's personality via a questionnaire then yields five values between 1 and 5, one for each of the traits.

Systematic technical measures for ensuring user privacy are a largely unaddressed aspect of psychoinformatics. When tracking data via mobile (crowd)sensing, deep insights about the user are possible, potentially risking his/her privacy, especially when dealing with psychological data. Depending on the region or specific use case of related apps in this field, certain privacy regulations have to be complied with. For example, the European Union implemented the General Data Protection Regulation (GDPR) in 2018 [71]. Deploying a psychoinformatical or a mobile crowdsensing app as a medical product needs certifications that might have additional specific requirements. Overall, when collecting mobile data, structured approaches in dealing with privacy have to be developed and implemented. However, in existing research, user privacy is often not considered at all [70], or no technical details are given [26, 18, 67, 22, 49, 38, 52, 83, 20, 60]. Furthermore, researchers conducting psychoinformatical studies face the problem that it is unclear to what extent demographics, education, and personality traits play a role when collecting data from users in daily life, which is crucial for study planning. While there is a variety of research on smartphone usage and personality traits [24, 26, 41, 40, 77, 76], there are still important knowledge gaps, for example, relating to the relationship between personality traits and the frequency of smartphone usage and usage session duration.

New knowledge about personality traits and smartphone usage helps deepen the understanding of human behavior. This, in turn, is crucial for psychological research, e.g., related to smartphone overuse [82, 51], and for a variety of fields in computer science and software development, especially when trying to understand a user base or tailoring services to customers' personalities. Mobile health (mHealth) applications benefit from personality information for the diagnosis or treatment of patients [68, 61, 87, 69, 39], e.g., giving individualized feedback for patients. In the field of recommender systems, understanding the user's personality helps in the calculation of recommendations and in understanding user behavior when interacting with recommendations [48, 3]. The importance of personality for the attitude toward advertising and mobile commerce is highlighted, for example, in [65, 84, 57]. Additionally, more knowledge about smartphone usage and personality traits serves as a vital building block toward the prediction of aspects of the user's personality without applying questionnaires [43].

1.2   Ubiquitous Social Networking

When looking into what people actually do with their smartphones, we observe a high focus on the interaction with other people by using messaging apps and social networking apps. Overall, 19.83% of time spent with the phone is spent on the messenger app WhatsApp alone, and 9.38% of time is spent on Facebook [63]. These two apps alone already account for almost a third of all smartphone usage. These observations reflect research describing social bonding as the "most crucial human motive" [73]. Psychological, as well as sociological, research can help explain when and how social bonding works. Besides the already introduced concept of *personality*, two additional concepts are of importance here: *homophily* and *propinquity*. Not only is personality connected to everyday behavior, but it also plays an integral part for social interactions and the forming of interpersonal relationships [21, 74]. *Homophily* describes the tendency of similar people to form meaningful bonds with each other. According to existing research, homophily structures any type of network [59]. *Propinquity* describes how being in physical proximity often is a deciding factor for interpersonal attraction [56, 32]. These research results from psychology and social sciences indicate that *similarity* and *proximity* are deciding factors to be taken into account when developing social networking applications.

Applications that deal with the interactions between users in proximity lie at the intersection of *ubiquitous computing* and *social networking services*. Sometimes, this research field is also referred to as *proximity(-based) mobile social networks* [42]. The most common use case in this field is the recommendation of new contacts in proximity for the incentivization of social interaction. Psychological principles of what factors actually play a role for forming interpersonal bonds have rarely been considered. By integrating psychoinformatics with ubiquitous social networking, potentially more meaningful recommendations for new contacts can be given. Furthermore, most works rely on manually entered profile data like lists of interests [2, 80, 23, 78, 79, 64]. Such manual processes make it tedious for the user to use such systems, and constraints apply, for example, with respect to the often predefined lists of interests to choose from. With a higher degree of automation, inhibition thresholds for using such system are lowered, and services for individual users or groups of users can be provided seamlessly.

1.3   Research Objectives

The overall research goal of this thesis is related to the understanding and well-being of people. Psychoinformatical research related to personality traits and

smartphone usage helps understand human behavior and can ultimately support
the treatment of patients, understanding users of software systems, and tailoring
individualized services. Ubiquitous social networking research serves as a tool for
the incentivization of social interaction between people. In order to achieve this
goal, first, in the field of psychoinformatics, we are researching the relationship
between smartphone usage and the user's personality. Second, based on existing
psychological research about the formation of interpersonal bonds, we research the
implications this has for social networking applications and develop prototypical
concepts and applications incorporating our findings. Thus, for the integration of
psychoinformatics with ubiquitous social networking, this thesis addresses two main
research questions:

Research Question A   What is the relationship between smartphone sensor and
usage data and the user's personality?

Research Question B   How to use insights from psychological research for the
design of social networking applications?

The links between the two research questions are threefold: (a) on a conceptual
level, we deal with the relationship between context data and psychological concepts
like personality; (b) on a technical level, the collection of data, mobile sensing,
serves as the basis; and (c) on a thematic level, privacy has to be a core focus when
addressing either question.

In order to answer Research Question A, we pose the following research tasks:

A-Task1   *Survey related work in the field of psychoinformatics.*
Based on the initial outline in this section, we need to deepen what *personality*
means, how it is assessed, and what the state of the art in psychoinformatical
research is. This survey includes fundamentals of psychology, and it focuses
on psychoinformatical studies dealing with smartphone data in relation to
psychological concepts of the user.

A-Task2   *Develop an app for conducting psychoinformatical research that takes
into account privacy awareness.*
In order to be able to analyze the relationship between smartphone data and
personality, we need an app that allows for the related data collection. This task
includes the design, implementation, and release of a mobile application that
can be used for studies in the field of psychoinformatics. In order to be used
for more research in the future, the app should provide objective measurements
of any data that are potentially indicative of psychological traits of the user,
i.e., all sensor data and usage statistics. Given the privacy concerns related
to psychoinformatical or mobile crowdsensing apps, a structured approach to
dealing with privacy has to be developed and implemented.

A-Task3   *Research what data users are willing to share with researchers and how
the sharing of data relates to user characteristics and personality traits.*
This task relates to the improvement of study planning of future psychoinformatical studies. The relationship between user characteristics like age, gender,

education, and personality traits and the data users are willing to share in mobile crowdsensing studies should be analyzed.

A-Task4   *Research the relationship between smartphone usage frequency, usage session duration, and personality traits of the user.*
The app from A-Task2 should be used to collect data from users, capturing their smartphone usage behavior. Based on these data, a concrete study should analyze the relationship between smartphone usage and personality. This serves two purposes: first, it is to demonstrate the functionality of the results of the previous tasks; second, the new knowledge helps in future research in this field, for example, related to smartphone overuse or to the understanding of user behavior.

In order to answer Research Question B, we pose the following research tasks:

B-Task1   *Survey related work in the field of ubiquitous social networking.*
Terms and fundamentals about psychological concepts like homophily and propinquity need to be reviewed, and the related work in the field of ubiquitous social networking should be surveyed.

B-Task2   *Develop a concept for the incentivization of social interaction, incorporating findings from research in psychology and social sciences.*
One of the main themes in ubiquitous social networking is the incentivization of social interaction between strangers. When developing a concept for this, the findings surveyed in B-Task1 should be incorporated.

B-Task3   *Develop a concept to assess the similarity of users in ubiquitous social networking scenarios.*
Psychology and social sciences show that the key concept structuring social networks is homophily. This concept translates into the similarity of users in social networking services. The goal of this task is the assessment of the similarity of users of a ubiquitous social networking service. User similarity should be assessed in an automated way in order to lower the inhibition threshold for using systems implementing it.

B-Task4   *Develop example ubiquitous social networking applications taking into account the results of the previous tasks.*
Based on the findings from the previous tasks, example applications should be developed. The goal is to show the functionality and feasibility of our core idea of integrating research results from psychology and social sciences into the design of social networking applications.

1.4   Contributions

In the following, we summarize the main contributions of this thesis for both research questions.

1.4.1   Contributions Relating to Research Question A

With TYDR—Track Your Daily Routine—we present our app for psychoinformatical research. TYDR tracks more smartphone data than most existing tools. TYDR is a publicly available app and will serve as a tool in further research projects in the future. With PM-MoDaC, our privacy model for apps related to mobile data collection, to the best of our knowledge, we are the first to provide concrete measures that researchers and software developers can implement for the protection of their users' privacy in mobile crowdsensing apps. By analyzing user demographics and personality traits in relation to willingness to share data with researchers, we help researchers of future psychoinformatical studies in study planning. We analyzed the relationship between smartphone usage frequency and session duration and user demographics and personality traits. To the best of our knowledge, we are the first to look at these aspects specifically. The resulting new knowledge benefits psychologists and developers of mobile systems.

In the following four paragraphs, we will describe these four main contributions relating to Research Question A in more detail:

TYDR: App for Psychoinformatical Research We developed, implemented, and evaluated a mobile system that enables the research of the relationship between smartphone sensor data and usage statistics and the users' personality. Our Android app TYDR tracks smartphone data and utilizes psychometric personality questionnaires. With TYDR, we track a larger variety of smartphone data than most similar existing apps, including metadata on notifications, photos taken, and music played back by the user. We optimized the tracking of sensor data by assessing the trade-off of the size of data and battery consumption and granularity of the stored information. Our user interface is designed to incentivize users to install the app and fill out questionnaires. TYDR processes and visualizes the tracked sensor and usage data as well as the results of the personality questionnaires. The ethics commissions of Technische Universität Berlin approved the use of TYDR for our psychoinformatical studies. We released TYDR on Google Play in October 2018 and registered 3,921 installations since then. TYDR is planned to be used in future projects in Germany and Austria for further psychoinformatical studies with thousands of users.

Privacy Model for Mobile Data Collection Apps Surveying the related psychoinformatical work, we saw a lack of a consistent approach to privacy awareness for apps relating to the mobile collection of data. To the best of our knowledge, we are the first to propose a full-scale, integrated privacy model. Our privacy model called PM-MoDaC consists of nine concrete measures to be taken to ensure the participants'/users' privacy. These measures include transparently informing the users about the data being collected, the anonymization of user data, and enabling an opt-out option. As part of our privacy model, we present a process for anonymized data storing while still being able to identify individual users who successfully completed a psychological study with the app. We present the implementation of all our privacy measures in TYDR. Our privacy model PM-MoDaC has already started to gain visibility, and other researchers started utilizing our approach for their research (e.g., [40]).

Analysis of Users' Willingness to Share Data with Researchers We collected data with TYDR over the course of a 2-month period and extensively evaluated our privacy model and which data users are willing to share. We found evidence that our users accept our proposed privacy model. Based on the data about granting TYDR all or no Android system permissions, we found evidence that younger users tend to be less willing to share their data (average age of 30 years compared to 35 years). We also observed that female users tend to be less willing to share data compared to male users. We did not find any evidence that education or personality traits are a factor related to data sharing. TYDR users score higher on the personality trait *openness to experience* than the average of the population, which we assume to be evidence that the type of app influences the user base it attracts in terms of average personality traits. We believe that our findings can help other researchers conducting similar research estimate the data to be expected.

Smartphone Usage Frequency and Usage Session Duration in Relation to Personality Traits We conducted a psychoinformatical study and investigated associations between personality traits and smartphone usage in daily life. Based on 11 months of data collection with TYDR, we analyzed 526 participants (mean age 34.57 years, SD = 12.85, 21% female) who provided data for 48 days, on average (SD = 63.2, range 2 to 304). We measured the Big Five personality traits (openness, conscientiousness, extraversion, agreeableness, neuroticism). Using hierarchical linear models, we analyzed associations between personality traits and (1) number of screen wakeups and (2) average session duration. On average, our users used their smartphone 72 times per day, for 3.7 minutes each time on average. We found that participants reached for their smartphone more frequently during weekdays with a shorter duration of usage compared to weekends. Younger people used their smartphones more often but with a shorter duration than older people. Female participants spent more time using smartphones per session than male participants. Extraversion and neuroticism were associated with more frequent usage of the smartphone per day, while conscientiousness was associated with a shorter mean session duration. Our results are helpful for the assessment and treatment of patients, for example, related to smartphone overuse. Additionally, our results help developers of mobile systems individualize services and understand their users' behavior.

1.4.2 Contributions Relating to Research Question B

With SimCon, we present a concept for the smartphone-mediated incentivization of social interaction between strangers. The concept is based on mobile sensing and can be implemented in a completely unobtrusive way without any necessary user interaction. To the best of our knowledge, we are the first to explicitly consider psychological research results in such a concept, as well as the first to base the concept on mobile sensing, enabling full automation and thus lowering the inhibition threshold for usage. SimCon is based on the similarity of users. We introduce metrics for similarity estimation in ubiquitous social networking scenarios utilizing device-to-device communication and probabilistic data structures. The most important metric we introduce is CBF-Dice, which is generalizable for the fast and space-efficient similarity estimation of any two multisets. With MobRec and GroupMusic, we present two highly automated example applications for ubiquitous social networks. MobRec is a platform for recommender systems that does not exhibit the typical lock-in effects of such systems. For MobRec, we implemented and evaluated unobtrusive device-to-device data exchange running in the background on off-the-shelf iOS and Android devices. GroupMusic is a group recommender system that plays back music according to the taste of currently present users. The architecture and example applications can serve as blueprints or best practices for researchers and developers in the fields of device-to-device computing, mobile software engineering, and recommender systems.

In the following four paragraphs, we will describe these four main contributions relating to Research Question B in more detail:

Concept for Contact Recommendations: SimCon While existing social networking services tend to connect people who know each other, people show a desire to also connect to yet unknown people in physical proximity. This is reflected in the fact that the most prevalent topic in the field of ubiquitous social networking is the incentivization of social interaction between strangers. Based on the concepts of personality, homophily, and propinquity, we designed a concept for contact recommendations. The idea is that the context data on the phone reflects personality as well as preferences, for example, with respect to visited locations, media played back, or structuring of daily life. Thus, similar context data implies similar personality and preferences. Our concept SimCon proposes to exchange such context data in a device-to-device manner and to calculate the similarity of the exchanged data in order to recommend similar users in proximity. SimCon is based on mobile sensing, so no data has to be entered manually by the user. The whole contact recommendation process can run fully automatically, lowering the inhibition threshold and potentially providing more meaningful recommendations.

Metrics for Similarity Estimation of Multisets Building on the previous main contribution, we proposed, developed, and evaluated metrics for estimating the similarity between two smartphone users in ubiquitous social networking settings. Our approach is generalizable for any other similarity estimation of frequencies represented as multisets. We developed our solution based on the example of similarity in musical taste. We showed that a single exchange of a probabilistic data structure between two devices can closely estimate the similarity of two users— without the need to contact a third-party server. We introduce metrics for fast and space-efficient approximation of the Dice coefficient of two multisets—based on the comparison of two Counting Bloom filters or two Count-Min Sketches. Our analysis shows that utilizing a single hash function minimizes the error when comparing these probabilistic data structures. The size that should be chosen for the data structure depends on the expected average number of unique input elements. In an experimental study, using real user data, we showed that a Counting Bloom Filter with a single hash function and a small length—twice the size of the average number of unique input elements—is sufficient to accurately estimate the similarity between two multisets representing the musical tastes of two users.

Mobile Platform for Decentralized Recommender Systems: MobRec As an example application for ubiquitous social networking services, we developed a mobile platform for decentralized recommender systems. The core concept is that everything runs on smartphones. Ratings and preferences are captured locally. During daily life, these ratings and preferences are exchanged with users in proximity in a device-to-device manner. Locally running recommender systems or third-party service providers can then recommend items based on own data and data received from users met before. We implemented the platform for off-the-shelf smartphones for both Android and iOS. We implemented device-to-device data exchange that can run in the background on both platforms. In our evaluation, we showed the feasibility of our approach. Our implementation can serve as a blueprint for future research and software development in this field.

Group Music Recommender System: GroupMusic We developed another example application for ubiquitous social networking services, for groups of users. We created the system GroupMusic that allows the generation and playback of group music playlists that are based on the musical taste of individual guests attending a meeting. In our architecture, we utilize automatically collected data on smartphones for the automatized generation of group music playlists. We follow the idea of utilizing context data in a preprocessing step to generate a group music profile for the recommendation process that generates a group music playlist. The whole process is almost fully automated, and the played back music automatically adapts to the users currently present. This contribution is especially relevant for researchers and developers of ubiquitous computing systems.

1.5 Outline

Reflecting the two overall research questions addressed, this thesis is divided into two main parts, each dealing with one of the research questions A and B.

Part I focuses on deepening the scientific understanding of people and their behavior and addresses research question A (*What is the relationship between smartphone sensor and usage data and the user's personality?*). We start by surveying related psychoinformatical work in Chap. 3 (A-Task1). We introduce our mobile system TYDR, which enables the research of the relationship between smartphone sensor data and usage statistics and the users' personality (A-Task2) in Chap. 4. This includes our privacy model PM-MoDaC and data analysis regarding what data users are willing to share with researchers in mobile crowdsensing scenarios (A-Task3). Based on data we collected in a study conducted with TYDR, we analyze smartphone usage frequency and duration in relation with personality traits in Chap. 5 (A-Task4).

Part II focuses on offering services based on psychological principles of social behavior. Here, we address research question B (*How to use insights from psychological research for the design of social networking applications?*). Addressing B-Task1 in Chap. 8, we extensively survey related work. Based on our findings, we develop a concept for the incentivization of social interaction called SimCon in Chap. 9 (B-Task2). Building on these results, we propose, develop, and evaluate metrics for estimating the similarity of two users in Chap. 10 (B-Task3). Chapters 11 and 12 serve as examples for ubiquitous social networking applications that we developed based on our previous results (B-Task4). In Chap. 11, we present our mobile platform for decentralized recommender systems MobRec. In Chap. 12, we present an example application GroupMusic for groups of users that plays back music according to the taste of users currently present.

In Chap. 14, we summarize our findings, and Chap. 15 gives an outlook for further research.

1.6   Publications

Parts of this thesis have previously been published in the following 11 papers. One received a best paper award and one a best paper runner-up award.

- F. Beierle, S. Göndör, and A. Küpper. "Towards a Three-Tiered Social Graph in Decentralized Online Social Networks". In: *Proc. 7th International Workshop on Hot Topics in Planet-Scale mObile Computing and Online Social neTworking (HotPOST)*. ACM, June 2015, pp. 1–6. DOI: 10.1145/2757513.2757517 [8] [**Best Paper Runner-up**]
- F. Beierle, K. Grunert, S. Göndör, and A. Küpper. "Privacy-Aware Social Music Playlist Generation". In: *Proc. 2016 IEEE International Conference on Communications (ICC)*. IEEE, May 2016, pp. 5650–5656. DOI: 10.1109/ICC.2016.7511602 [14]
- F. Beierle, K. Grunert, S. Göndör, and V. Schlüter. "Towards Psychometrics-Based Friend Recommendations in Social Networking Services". In: *2017 IEEE International Conference on AI & Mobile Services (AIMS)*. IEEE, June 2017, pp. 105–108. DOI: 10.1109/AIMS.2017.22 [15]
- F. Beierle, V. T. Tran, M. Allemand, P. Neff, W. Schlee, T. Probst, R. Pryss, and J. Zimmermann. "TYDR—Track Your Daily Routine. Android App for Tracking Smartphone Sensor and Usage Data". In: *2018 IEEE/ACM 5th International Conference on Mobile Software Engineering and Systems (MOBILESoft)*. ACM, 2018, pp. 72–75. DOI: 10.1145/3197231.3197235 [16]

- F. Beierle. "Do You Like What I Like? Similarity Estimation in Proximity-Based Mobile Social Networks". In: *2018 17th IEEE International Conference On Trust, Security And Privacy In Computing And Communications (TrustCom)*. IEEE, Aug. 2018, pp. 1040–1047. DOI: `10.1109/TrustCom/BigDataSE.2018.00146` [4]
- F. Beierle, V. T. Tran, M. Allemand, P. Neff, W. Schlee, T. Probst, R. Pryss, and J. Zimmermann. "Context Data Categories and Privacy Model for Mobile Data Collection Apps". In: *Procedia Computer Science*. The 15th International Conference on Mobile Systems and Pervasive Computing (MobiSPC) 134 (2018), pp. 18–25. DOI: `10.1016/j.procs.2018.07.139` [12] **[Best Paper]**
- F. Beierle and T. Eichinger. "Collaborating with Users in Proximity for Decentralized Mobile Recommender Systems". In: *Proc. IEEE 16th International Conference on Ubiquitous Intelligence and Computing (UIC)*. IEEE, June 2019, pp. 1192–1197. DOI: `10.1109/SmartWorld-UIC-ATC-SCALCOM-IOP-SCI.2019.00222` [7]
- T. Eichinger, F. Beierle, R. Papke, L. Rebscher, H. C. Tran, and M. Trzeciak. "On Gossip-Based Information Dissemination in Pervasive Recommender Systems". In: *Prod. 13th ACM Conference on Recommender Systems (RecSys)*. ACM, Sept. 2019, pp. 442–446. DOI: `10.1145/3298689.3347067` [31]
- F. Beierle, V. T. Tran, M. Allemand, P. Neff, W. Schlee, T. Probst, J. Zimmermann, and R. Pryss. "What Data Are Smartphone Users Willing to Share with Researchers?" In: *Journal of Ambient Intelligence and Humanized Computing* 11 (2020), pp. 2277–2289. DOI: `10.1007/s12652-019-01355-6` [17]
- F. Beierle, T. Probst, M. Allemand, J. Zimmermann, R. Pryss, P. Neff, W. Schlee, S. Stieger, and S. Budimir. "Frequency and Duration of Daily Smartphone Usage in Relation to Personality Traits". In: *Digital Psychology* 1.1 (2020), pp. 20–28. DOI: `10.24989/dp.v1i1.1821` [13]
- F. Beierle and S. Egger. "MobRec—Mobile Platform for Decentralized Recommender Systems". In: *IEEE Access* 8 (2020), pp. 185311–185329. DOI: `10.1109/ACCESS.2020.3029319` [6]

Some parts of this thesis are appearing in the following book chapter:

- F. Beierle, S. C. Matz, and M. Allemand. "Smartphone Sensing in Personality Science". In: *Mobile Sensing in Psychology: Methods and Applications*. Ed. by M. R. Mehl, C. Wrzus, M. Eid, G. Harari, and U. E. Priemer. New York City, NY, USA: Guilford Press, 2021 (to appear) [9]

While I was working on this thesis, I also contributed to 17 additional publications which are only loosely related to this thesis. Two publications are related to attribute-based encryption in mobile scenarios [85, 27]. Three publications are related to the global reachability of users registered with different communication service providers. We utilized a decentralized directory supporting the storing of global reachability information [47, 33, 46]. In [37, 36, 35], we followed a similar approach. Here, we proposed solutions for a federation of online social networking services—including a decentralized global reachability registry and mechanisms for the migration of user profiles to different service providers. In the field of recommender systems, we built and evaluated a system for conference recommendations based on different relations between authors [10]. The publications [5] and [11] deal with the psychological concept of *choice overload* or *overchoice*, the tendency of people to not make any decision when faced with too many options. We analyzed this effect in related-article recommendations in digital libraries. In [30], we developed a method to calculate a latent similarity of two users based on their texting data. This is related to B-Task3 (*Develop a concept to assess the similarity of users in ubiquitous social networking scenarios.*). Additional publications are related to software architectures for data analysis [86, 28, 29] and to blockchain [54, 53]:

- S. Zickau, F. Beierle, and I. Denisow. "Securing Mobile Cloud Data with Personalized Attribute-Based Meta Information". In: *Proc. 3rd IEEE International Conference on Mobile Cloud Computing, Services, and Engineering (MobileCloud)*. IEEE, Mar. 2015, pp. 205–210. DOI: 10.1109/MobileCloud.2015.14 [85]
- I. Denisow, S. Zickau, F. Beierle, and A. Küpper. "Dynamic Location Information in Attribute-Based Encryption Schemes". In: *Proc. 9th International Conference on Next Generation Mobile Applications, Services and Technologies (NGMAST)*. IEEE, 2015, pp. 240–247. DOI: 10.1109/NGMAST.2015.63 [27]
- I. T. Javed, R. Copeland, N. Crespi, F. Beierle, S. Göndör, A. Küpper, M. Emmelmann, A. Corici, K. Corre, J.-M. Crom, A. Bouabdallah, F. Oberle, I. Friese, A. Caldeira, G. Dias, R. Chaves, and N. Santo. "Global Identity and Reachability Framework for Interoperable P2P Communication Services". In: *Proc. 2016 Conference on Innovations in Clouds, Internet and Networks (ICIN)*. IFIP, 2016, pp. 59–66 [47]
- I. Friese, R. Copeland, S. Göndör, F. Beierle, A. Küpper, R. L. Pereira, and J.-M. Crom. "Cross-Domain Discovery of Communication Peers: Identity Mapping and Discovery Services (IMaDS)". In: *Proc. 2017 European Conference on Networks and Communications (EuCNC)*. IEEE, June 2017, pp. 1–6. DOI: 10.1109/EuCNC.2017.7980642 [33]
- I. T. Javed, R. Copeland, N. Crespi, M. Emmelmann, A. Corici, A. Bouabdallah, T. Zhang, S. El Jaouhari, F. Beierle, S. Göndör, A. Küpper, K. Corre, J.-M. Crom, F. Oberle, I. Friese, A. Caldeira, G. Dias, N. Santos, R. Chaves, and R. L. Pereira. "Cross-Domain Identity and Discovery Framework for Web Calling Services". In: *Annals of Telecommunications* 72.7 (Aug. 2017), pp. 459–468. DOI: 10.1007/s12243-017-0587-2 [46]
- S. Göndör, F. Beierle, E. Kücükbayraktar, H. Hebbo, S. Sharhan, and A. Küpper. "Towards Migration of User Profiles in the SONIC Online Social Network Federation". In: *Proc. International Multi-Conference on Computing in the Global Information Technology (ICCGI)*. IARIA, 2015, pp. 1–5 [37]
- S. Göndör, F. Beierle, S. Sharhan, H. Hebbo, E. Kücükbayraktar, and A. Küpper. "SONIC: Bridging the Gap between Different Online Social Net-

work Platforms". In: *Proc. 2015 IEEE International Conference on Smart City/SocialCom/SustainCom (SmartCity)*. IEEE, Dec. 2015, pp. 399–406. DOI: 10.1109/SmartCity.2015.104 [36]

- S. Göndör, F. Beierle, S. Sharhan, and A. Küpper. "Distributed and Domain-Independent Identity Management for User Profiles in the SONIC Online Social Network Federation". In: *International Conference on Computational Social Networks (CSoNet)*. Ed. by H. T. Nguyen and V. Snasel. Vol. 9795. LNCS. Springer, 2016, pp. 226–238. DOI: 10.1007/978-3-319-42345-6_20 [35]

- F. Beierle, J. Tan, and K. Grunert. "Analyzing Social Relations for Recommending Academic Conferences". In: *Proc. 8th ACM International Workshop on Hot Topics in Planet-Scale mObile Computing and Online Social neTworking (HotPOST)*. ACM, 2016, pp. 37–42. DOI: 10.1145/2944789.2944871 [10]

- F. Beierle, A. Aizawa, and J. Beel. "Exploring Choice Overload in Related-Article Recommendations in Digital Libraries". In: *Proc. 5th International Workshop on Bibliometric-Enhanced Information Retrieval (BIR)*. Vol. 1823. CEUR Workshop Proceedings. CEUR-WS, 2017, pp. 51–61 [5]

- T. Eichinger, F. Beierle, S. U. Khan, R. Middelanis, S. Veeraraghavan, and S. Tabibzadeh. "Affinity: A System for Latent User Similarity Comparison on Texting Data". In: *Proc. 2019 IEEE International Conference on Communications (ICC)*. IEEE, May 2019, pp. 1–7. DOI: 10.1109/ICC.2019.8761051 [30]

- E. Zielinski, J. Schulz-Zander, M. Zimmermann, C. Schellenberger, A. Ramirez, F. Zeiger, M. Mormul, F. Hetzelt, F. Beierle, H. Klaus, H. Ruckstuhl, and A. Artemenko. "Secure Real-Time Communication and Computing Infrastructure for Industry 4.0—Challenges and Opportunities". In: *Proc. 2019 International Conference on Networked Systems (NetSys)*. IEEE, Mar. 2019, pp. 1–6. DOI: 10.1109/NetSys.2019.8854499 [86]

- H. Dinh-Tuan, F. Beierle, and S. Rodriguez Garzon. "MAIA: A Microservices-Based Architecture for Industrial Data Analytics". In: *Proc. 2019 IEEE International Conference on Industrial Cyber Physical Systems (ICPS)*. IEEE, May 2019, pp. 23–30. DOI: 10.1109/ICPHYS.2019.8780345 [28]

- Z. A. Lux, F. Beierle, S. Zickau, and S. Göndör. "Full-Text Search for Verifiable Credential Metadata on Distributed Ledgers". In: *Proc. 2019 Sixth International Conference on Internet of Things: Systems, Management and Security (IOTSMS)*. IEEE, Oct. 2019, pp. 519–528. DOI: 10.1109/IOTSMS48152.2019.8939249 [54]

- F. Beierle, A. Aizawa, A. Collins, and J. Beel. "Choice Overload and Recommendation Effectiveness in Related-Article Recommendations". In: *International Journal on Digital Libraries* 21.3 (Sept. 2020), pp. 231–246. DOI: 10.1007/s00799-019-00270-7 [11]

- Z. A. Lux, D. Thatmann, S. Zickau, and F. Beierle. "Distributed-Ledger-Based Authentication with Decentralized Identifiers and Verifiable Credentials". In: *2020 2nd Conference on Blockchain Research Applications for Innova-*

tive Networks and Services (BRAINS). IEEE, Sept. 2020, pp. 71–78. DOI: `10.1109/BRAINS49436.2020.9223292` [53]
- H. Dinh-Tuan, M. Mora-Martinez, F. Beierle, and S. Rodriguez Garzon. "Development Frameworks for Microservice-based Applications: Evaluation and Comparison". In: *Proc. 2020 European Symposium on Software Engineering (ESSE)*. ACM, Nov. 2020, pp. 12–20. DOI: `10.1145/3393822.3432339` (in print) [29]

References

1. H. Baumeister, C. Montag (eds.), *Digital Phenotyping and Mobile Sensing: New Developments in Psychoinformatics*. Studies in Neuroscience, Psychology and Behavioral Economics (Springer International Publishing, 2019). https://doi.org/10.1007/978-3-030-31620-4
2. A. Beach, M. Gartrell, S. Akkala, J. Elston, J. Kelley, K. Nishimoto, B. Ray, S. Razgulin, K. Sundaresan, B. Surendar, M. Terada, R. Han, WhozThat? Evolving an ecosystem for context-aware mobile social networks. IEEE Netw. **22**(4), 50–55 (2008)
3. J. Beel, C. Breitinger, S. Langer, A. Lommatzsch, B. Gipp, Towards reproducibility in recommender-systems research. User Model. User-Adapted Interaction **26**(1), 69–101 (2016). https://doi.org/10.1007/s11257-016-9174-x
4. F. Beierle, Do you like what I like? Similarity estimation in proximity-based mobile social networks, in *2018 17th IEEE International Conference on Trust, Security and Privacy in Computing and Communications (TrustCom)* (IEEE, 2018), pp. 1040–1047. https://doi.org/10.1109/TrustCom/BigDataSE.2018.00146
5. F. Beierle, A. Aizawa, J. Beel, Exploring choice overload in related-article recommendations in digital libraries, in *Proceedings of 5th International Workshop on Bibliometric-Enhanced Information Retrieval (BIR)*, vol. 1823. CEUR Workshop Proceedings (CEUR-WS, 2017), pp. 51–61
6. F. Beierle, S. Egger, MobRec—mobile platform for decentralized recommender systems. IEEE Access **8**, 185311–185329 (2020). https://doi.org/10.1109/ACCESS.2020.3029319
7. F. Beierle, T. Eichinger, Collaborating with users in proximity for decentralized mobile recommender systems, in *Proceedings of IEEE 16th International Conference on Ubiquitous Intelligence and Computing (UIC)* (IEEE, 2019), pp. 1192–1197. https://doi.org/10.1109/SmartWorld-UIC-ATC-SCALCOM-IOP-SCI.2019.00222
8. F. Beierle, S. Göndör, A. Küpper, Towards a three-tiered social graph in decentralized online social networks, in *Proceedings of 7th International Workshop on Hot Topics in Planet-Scale mObile Computing and Online Social neTworking (HotPOST)* (ACM, 2015), pp. 1–6. https://doi.org/10.1145/2757513.2757517
9. F. Beierle, S.C. Matz, M. Allemand, Smartphone sensing in personality science, in *Mobile Sensing in Psychology: Methods and Applications*, ed. by M.R. Mehl, C. Wrzus, M. Eid, G. Harari, U.E. Priemer (Guilford Press, New York City, 2021)
10. F. Beierle, J. Tan, K. Grunert, Analyzing social relations for recommending academic conferences, in *Proceedings of 8th ACM International Workshop on Hot Topics in Planet-Scale Mobile Computing and Online Social Networking (HotPOST)* (ACM, 2016), pp. 37–42. https://doi.org/10.1145/2944789.2944871
11. F. Beierle, A. Aizawa, A. Collins, J. Beel, Choice overload and recommendation effectiveness in related-article recommendations. Int. J. Digit. Libr. **21**(3), 231–246 (2020). https://doi.org/10.1007/s00799-019-00270-7
12. F. Beierle, V.T. Tran, M. Allemand, P. Neff, W. Schlee, T. Probst, R. Pryss, J. Zimmermann, Context data categories and privacy model for mobile data collection apps, in *Procedia Computer Science* **134**, 18–25 (2018). The 15th International Conference on Mobile Systems and Pervasive Computing (MobiSPC). https://doi.org/10.1016/j.procs.2018.07.139

13. F. Beierle, T. Probst, M. Allemand, J. Zimmermann, R. Pryss, P. Neff, W. Schlee, S. Stieger, S. Budimir, Frequency and duration of daily smartphone usage in relation to personality traits. *Digital Psychology* **1**(1), 20–28 (2020). https://doi.org/10.24989/dp.v1i1.1821

14. F. Beierle, K. Grunert, S. Göndör, A. Küpper, Privacy-aware social music playlist generation, in *Procedings of 2016 IEEE International Conference on Communications (ICC)* (IEEE, 2016), pp. 5650–5656. https://doi.org/10.1109/ICC.2016.7511602

15. F. Beierle, K. Grunert, S. Göndör, V. Schlüter, Towards psychometrics-based friend recommendations in social networking services, in *2017 IEEE International Conference on AI & Mobile Services (AIMS)* (IEEE, 2017), pp. 105–108. https://doi.org/10.1109/AIMS.2017.22

16. F. Beierle, V.T. Tran, M. Allemand, P. Neff, W. Schlee, T. Probst, R. Pryss, J. Zimmermann, TYDR—track your daily routine. Android app for tracking smartphone sensor and usage data, in *2018 IEEE/ACM 5th International Conference on Mobile Software Engineering and Systems (MOBILESoft)* (ACM, 2018), pp. 72–75. https://doi.org/10.1145/3197231.3197235

17. F. Beierle, V.T. Tran, M. Allemand, P. Neff, W. Schlee, T. Probst, J. Zimmermann, R. Pryss, What data are smartphone users willing to share with researchers? J. Ambient Intell. Hum. Comput. **11**, 2277–2289 (2020). https://doi.org/10.1007/s12652-019-01355-6

18. A. Bogomolov, B. Lepri, M. Ferron, F. Pianesi, A.S. Pentland, Daily stress recognition from mobile phone data, weather conditions and individual traits, in *Proceedings of the 22nd ACM International Conference on Multimedia*, MM'14 (ACM, 2014), pp. 477–486. https://doi.org/10.1145/2647868.2654933

19. D.E. Boubiche, M. Imran, A. Maqsood, M. Shoaib, Mobile crowd sensing—taxonomy, applications, challenges, and solutions. Comput. Hum. Behav. **101**, 352–370 (2019). https://doi.org/10.1016/j.chb.2018.10.028

20. M. Boukhechba, Y. Huang, P. Chow, K. Fua, B.A. Teachman, L.E. Barnes, Monitoring social anxiety from mobility and communication patterns, in *Proceedings of the 2017 ACM International Joint Conference on Pervasive and Ubiquitous Computing and Proceedings of the 2017 ACM International Symposium on Wearable Computers* (UbiComp'17) (ACM, 2017), pp. 749–753. https://doi.org/10.1145/3123024.3125607

21. S. Burgess, E. Sanderson, M. Umaña-Aponte. *School Ties: An Analysis of Homophily in an Adolescent Friendship Network*. CMPO Working Paper Series 11/267. Centre for Market and Public Organisation (2011)

22. S. Butt, J.G. Phillips, Personality and self reported mobile phone use. Comput. Hum. Behav. Part Special Issue: Cognition and Exploratory Learning in Digital Age **24**(2), 346–360 (2008). https://doi.org/10.1016/j.chb.2007.01.019

23. A.C. Champion, Z. Yang, B. Zhang, J. Dai, D. Xuan, D. Li, E-SmallTalker: a distributed mobile system for social networking in physical proximity. IEEE Trans. Parallel Distrib. Syst. **24**(8), 1535–1545 (2013). https://doi.org/10.1109/TPDS.2012.251

24. G. Chittaranjan, J. Blom, D. Gatica-Perez, Mining large-scale smartphone data for personality studies. Pers. Ubiquit. Comput. **17**(3), 433–450 (2013). https://doi.org/10.1007/s00779-011-0490-1

25. P.J. Corr, G. Matthews, The Cambridge handbook of personality psychology, in *The Cambridge Handbook of Personality Psychology*, ed. by P.J. Corr, G. Matthews (Cambridge University Press, 2009), pp. xxii–xlii. https://doi.org/10.1017/CBO9780511596544.002

26. Y.-A. de Montjoye, J. Quoidbach, F. Robic, A.S. Pentland, Predicting personality using novel mobile phone-based metrics, in *Social Computing, Behavioral-Cultural Modeling and Prediction*, ed. by A.M. Greenberg, W.G. Kennedy, N.D. Bos. Lecture Notes in Computer Science (Springer, 2013), pp. 48–55. https://doi.org/10.1007/978-3-642-37210-0_6

27. I. Denisow, S. Zickau, F. Beierle, A. Küpper, Dynamic location information in attribute-based encryption schemes, in *Proceedings of 9th International Conference on Next Generation Mobile Applications, Services and Technologies (NGMAST)* (IEEE, 2015), pp. 240–247. https://doi.org/10.1109/NGMAST.2015.63

28. H. Dinh-Tuan, F. Beierle, S. Rodriguez Garzon, MAIA: a microservices-based architecture for industrial data analytics, in *Proceedings of 2019 IEEE International Conference on Industrial Cyber Physical Systems (ICPS)* (IEEE, 2019), pp. 23–30. https://doi.org/10.1109/ICPHYS. 2019.8780345
29. H. Dinh-Tuan, M. Mora-Martinez, F. Beierle, S. Rodriguez Garzon, Development frameworks for microservice-based applications: evaluation and comparison, in *Proceedings of 2020 European Symposium on Software Engineering (ESSE)* (ACM, 2020), pp. 12–20. https://doi. org/10.1145/3393822.3432339
30. T. Eichinger, F. Beierle, S.U. Khan, R. Middelanis, S. Veeraraghavan, S. Tabibzadeh, Affinity: a system for latent user similarity comparison on texting data, in *Proceedings of 2019 IEEE International Conference on Communications (ICC)* (IEEE, 2019), pp. 1–7. https://doi.org/10. 1109/ICC.2019.8761051
31. T. Eichinger, F. Beierle, R. Papke, L. Rebscher, H.C. Tran, M. Trzeciak, On gossip-based information dissemination in pervasive recommender systems, in *Proceedings of 13th ACM Conference on Recommender Systems (RecSys)* (ACM, 2019), pp. 442–446. https://doi.org/10. 1145/3298689.3347067
32. L. Festinger, S. Schachter, K. Back, The spatial ecology of group formation, in *Social Pressures in Informal Groups* (Harper, New York, 1950), pp. 141–161
33. I. Friese, R. Copeland, S. Göndör, F. Beierle, A. Küpper, R.L. Pereira, J.-M. Crom, Cross-domain discovery of communication peers: identity mapping and discovery services (IMaDS), in *Proceedings of 2017 European Conference on Networks and Communications (EuCNC)* (IEEE, 2017), pp. 1–6. https://doi.org/10.1109/EuCNC.2017.7980642
34. R.K. Ganti, F. Ye, H. Lei, Mobile crowdsensing: current state and future challenges. IEEE Commun. Mag. **49**(11), 32–39 (2011). https://doi.org/10.1109/MCOM.2011.6069707
35. S. Göndör, F. Beierle, S. Sharhan, A. Küpper, Distributed and domain-independent identity management for user profiles in the SONIC online social network federation, in *International Conference on Computational Social Networks (CSoNet)*, vol. 9795, ed. by H.T. Nguyen, V. Snasel. LNCS (Springer, 2016), pp. 226–238. https://doi.org/10.1007/978-3-319-42345-6_20
36. S. Göndör, F. Beierle, S. Sharhan, H. Hebbo, E. Kücükbayraktar, A. Küpper, SONIC: bridging the gap between different online social network platforms, in *Proceedings of 2015 IEEE International Conference on Smart City/SocialCom/SustainCom (SmartCity)* (IEEE, 2015), pp. 399–406. https://doi.org/10.1109/SmartCity.2015.104
37. S. Göndör, F. Beierle, E. Kücükbayraktar, H. Hebbo, S. Sharhan, A. Küpper, Towards migration of user profiles in the SONIC online social network federation, in *Proceedings of International Multi-Conference on Computing in the Global Information Technology (ICCGI)* (IARIA, 2015), pp. 1–5
38. T. Grover, G. Mark, Digital footprints: predicting personality from temporal patterns of technology use, in *Proceedings of the 2017 ACM International Joint Conference on Pervasive and Ubiquitous Computing and Proceedings of the 2017 ACM International Symposium on Wearable Computers* (UbiComp'17) (ACM, 2017), pp. 41–44. https://doi.org/10.1145/ 3123024.3123139
39. S. Halko, J.A. Kientz, Personality and persuasive technology: an exploratory study on health-promoting mobile applications, in *Persuasive Technology*, vol. 6137, ed. by D. Hutchison, T. Kanade, J. Kittler, J.M. Kleinberg, F. Mattern, J.C. Mitchell, M. Naor, O. Nierstrasz, C. Pandu Rangan, B. Steffen, M. Sudan, D. Terzopoulos, D. Tygar, M.Y. Vardi, G. Weikum, T. Ploug, P. Hasle, H. Oinas-Kukkonen (Springer, Berlin/Heidelberg, 2010), pp. 150–161. https://doi. org/10.1007/978-3-642-13226-1_16
40. G.M. Harari, S.R. Müller, C. Stachl, R. Wang, W. Wang, M. Bühner, P.J. Rentfrow, A.T. Campbell, S.D. Gosling, Sensing sociability: individual differences in young adults' conversation, calling, texting, and app use behaviors in daily life. J. Pers. Soc. Psychol. (2019). https://doi. org/10.1037/pspp0000245
41. G.M. Harari, N.D. Lane, R. Wang, B.S. Crosier, A.T. Campbell, S.D. Gosling, Using smartphones to collect behavioral data in psychological science: opportunities, practical considerations, and challenges. Perspect. Psychol. Sci. **11**(6), 838–854 (2016). https://doi.org/ 10.1177/1745691616650285

42. W. He, Y. Huang, K. Nahrstedt, B. Wu, Message propagation in ad-hoc-based proximity mobile social networks, in *2010 8th IEEE International Conference on Pervasive Computing and Communications Workshops (PERCOM Workshops)* (2010), pp. 141–146. https://doi.org/10.1109/PERCOMW.2010.5470617

43. J. Hinds, A. Joinson, Human and computer personality prediction from digital footprints. Curr. Dir. Psychol. Sci. **28**(2), 204–211 (2019). https://doi.org/10.1177/0963721419827849

44. J.B. Hirsh, C.G. DeYoung, J.B. Peterson, Metatraits of the big five differentially predict engagement and restraint of behavior. J. Pers. **77**(4), 1085–1102 (2009). https://doi.org/10.1111/j.1467-6494.2009.00575.x

45. T.R. Insel, Digital phenotyping: technology for a new science of behavior. JAMA **318**(13), 1215–1216 (2017). https://doi.org/10.1001/jama.2017.11295

46. I.T. Javed, R. Copeland, N. Crespi, M. Emmelmann, A. Corici, A. Bouabdallah, T. Zhang, S. El Jaouhari, F. Beierle, S. Göndör, A. Küpper, K. Corre, J.-M. Crom, F. Oberle, I. Friese, A. Caldeira, G. Dias, N. Santos, R. Chaves, R.L. Pereira, Cross-domain identity and discovery framework for web calling services. Ann. Telecommun. **72**(7), 459–468 (2017). https://doi.org/10.1007/s12243-017-0587-2

47. I.T. Javed, R. Copeland, N. Crespi, F. Beierle, S. Göndör, A. Küpper, M. Emmelmann, A. Corici, K. Corre, J.-M. Crom, A. Bouabdallah, F. Oberle, I. Friese, A. Caldeira, G. Dias, R. Chaves, N. Santo, Global identity and reachability framework for interoperable P2P communication services, in *Proceedings of 2016 Conference on Innovations in Clouds, Internet and Networks (ICIN)* (IFIP, 2016), pp. 59–66

48. R.P. Karumur, T.T. Nguyen, J.A. Konstan, Personality user preferences and behavior in recommender systems, in *Information Systems Frontiers* (2017), pp. 1–25. https://doi.org/10.1007/s10796-017-9800-0

49. S.Y. Kim, H.J. Koo, H.Y. Song, A study on estimation of human personality from location visiting preference. J. Ambient Intell. Hum. Comput. **9**(3), 629–642 (2018). https://doi.org/10.1007/s12652-017-0459-7

50. R. Kraft, W. Schlee, M. Stach, M. Reichert, B. Langguth, H. Baumeister, T. Probst, R. Hannemann, R. Pryss, Combining mobile crowdsensing and ecological momentary assessments in the healthcare domain. Front. Neurosci. **14**(164) (2020). https://doi.org/10.3389/fnins.2020.00164

51. K. Lanaj, R.E. Johnson, C.M. Barnes, Beginning the workday yet already depleted? Consequences of late-night smartphone use and sleep. Organ. Behav. Hum. Decis. Process. **124**(1), 11–23 (2014). https://doi.org/10.1016/j.obhdp.2014.01.001

52. W. Lane, C. Manner, The influence of personality traits on mobile phone application preferences. J. Econ. Behav. Stud. **4**(5), 252–260 (2012)

53. Z.A. Lux, D. Thatmann, S. Zickau, F. Beierle, Distributed-ledger-based authentication with decentralized identifiers and verifiable credentials, in *2020 2nd Conference on Blockchain Research Applications for Innovative Networks and Services (BRAINS)* (IEEE, 2020), pp. 71–78. https://doi.org/10.1109/BRAINS49436.2020.9223292

54. Z.A. Lux, F. Beierle, S. Zickau, S. Göndör, Full-text search for verifiable credential metadata on distributed ledgers, in *Proceedings of 2019 Sixth International Conference on Internet of Things: Systems, Management and Security (IOTSMS)* (IEEE, 2019), pp. 519–528. https://doi.org/10.1109/IOTSMS48152.2019.8939249

55. A. Markowetz, K. Błaszkiewicz, C. Montag, C. Switala, T.E. Schlaepfer, Psycho-informatics: big data shaping modern psychometrics. Med. Hypotheses **82**(4), 405–411 (2014). https://doi.org/10.1016/j.mehy.2013.11.030

56. D.M. Marvin, Occupational propinquity as a factor in marriage selection. Q. Publ. Am. Stat. Assoc. **16**(123), 131–150 (1918). https://doi.org/10.1080/15225445.1918.10503750

57. S.C. Matz, M. Kosinski, G. Nave, D.J. Stillwell, Psychological targeting as an effective approach to digital mass persuasion. Proc. Natl. Acad. Sci. **114**(48), 12714–12719 (2017). https://doi.org/10.1073/pnas.1710966114

58. R.R. McCrae, O.P. John, An introduction to the five-factor model and its applications. J. Pers. **60**(2), 175–215 (1992)

59. M. McPherson, L. Smith-Lovin, J.M. Cook, Birds of a feather: homophily in social networks. Annu. Rev. Soc. **27**, 415–444 (2001)
60. A. Mehrotra, R. Hendley, M. Musolesi, Towards multi-modal anticipatory monitoring of depressive states through the analysis of human-smartphone interaction, in *Proceedings of the 2016 ACM International Joint Conference on Pervasive and Ubiquitous Computing: Adjunct* (UbiComp'16) (Association for Computing Machinery, 2016), pp. 1132–1138. https://doi.org/10.1145/2968219.2968299
61. M.J. Roche, A.L. Pincus, A.L. Rebar, D.E. Conroy, N. Ram, Enriching psychological assessment using a person-specific analysis of interpersonal processes in daily life. Assessment **21**(5), 515–528 (2014). https://doi.org/10.1177/1073191114540320
62. C. Montag, É. Duke, A. Markowetz, Toward psychoinformatics: computer science meets psychology. Comput. Math. Methods Med. **2016**, 1–10 (2016). https://doi.org/10.1155/2016/2983685
63. C. Montag, H. Baumeister, C. Kannen, R. Sariyska, E.-M. Meßner, M. Brand, Concept, possibilities and pilot-testing of a new smartphone application for the social and life sciences to study human behavior including validation data from personality psychology. J. Multidiscip. Sci. J. **2**(2), 102–115 (2019). https://doi.org/10.3390/j2020008
64. J. Müller, C. Anneser, M. Sandstede, L. Rieger, A. Alhomssi, F. Schwarzmeier, B. Bittner, I. Aslan, E. André, Honeypot: a socializing app to promote train commuters' wellbeing, in *Proceedings of the 17th International Conference on Mobile and Ubiquitous Multimedia* (MUM 2018) (ACM, 2018), pp. 103–108. https://doi.org/10.1145/3282894.3282901
65. S.D. Myers, S. Sen, A. Alexandrov, The moderating effect of personality traits on attitudes toward advertisements: a contingency framework. Manag. Mark. **5**(3), 3–20 (2010)
66. D.J. Ozer, V. Benet-Martínez, Personality and the prediction of consequential outcomes. Annu. Rev. Psychol. **57**(1), 401–421 (2005). https://doi.org/10.1146/annurev.psych.57.102904.190127
67. J.G. Phillips, S. Butt, A. Blaszczynski, Personality and self-reported use of mobile phones for games. CyberPsychol. Behav. **9**(6), 753–758 (2006). https://doi.org/10.1089/cpb.2006.9.753
68. R. Pryss, M. Reichert, B. Langguth, W. Schlee, Mobile crowd sensing services for tinnitus assessment, therapy, and research, in *2015 IEEE International Conference on Mobile Services (MS)* (IEEE, 2015), pp. 352–359. https://doi.org/10.1109/MobServ.2015.55
69. R. Pryss, T. Probst,W. Schlee, J. Schobel, B. Langguth, P. Neff, M. Spiliopoulou, M. Reichert, Prospective crowdsensing versus retrospective ratings of tinnitus variability and tinnitus–stress associations based on the TrackYourTinnitus Mobile Platform. Int. J. Data Sci. Anal. **8**(4), 327–338 (2019). https://doi.org/10.1007/s41060-018-0111-4
70. K.K. Rachuri, M. Musolesi, C. Mascolo, P.J. Rentfrow, C. Longworth, A. Aucinas, EmotionSense: a mobile phones based adaptive platform for experimental social psychology research, in *Proceedings of the 12th ACM International Conference on Ubiquitous Computing* (UbiComp'10) (ACM, 2010), pp. 281–290. https://doi.org/10.1145/1864349.1864393
71. *Regulation (EU) 2016/679 of the European Parliament and of the Council of 27 April 2016 on the Protection of Natural Persons with Regard to the Processing of Personal Data and on the Free Movement of Such Data, and Repealing Directive 95/46/EC (General Data Protection Regulation) (Text with EEA Relevance)* (2016)
72. P.J. Rentfrow, S.D. Gosling, The Do Re Mi's of everyday life: the structure and personality correlates of music preferences. J. Pers. Soc. Psychol. **84**(6), 1236–1256 (2003). https://doi.org/10.1037/0022-3514.84.6.1236
73. T.J. Scheff, *Microsociology: Discourse, Emotion, and Social Structure* (University of Chicago Press, 1990)
74. M. Selfhout, W. Burk, S. Branje, J. Denissen, M. Van Aken, W. Meeus, Emerging late adolescent friendship networks and big five personality traits: a social network approach. J. Pers. **78**(2), 509–538 (2010). https://doi.org/10.1111/j.1467-6494.2010.00625.x
75. R.S. Sharma, Clothing behaviour, personality and values: a correlational study. Psychol. Stud. **25**(2), 137–142 (1980)

76. C. Stachl, Q. Au, R. Schoedel, D. Buschek, S. Völkel, T. Schuwerk, M. Oldemeier, T. Ullmann, H. Hussmann, B. Bischl, M. Bühner, *Behavioral Patterns in Smartphone Usage Predict Big Five Personality Traits*. Preprint. PsyArXiv (2019). https://doi.org/10.31234/osf.io/ks4vd

77. C. Stachl, S. Hilbert, J.-Q. Au, D. Buschek, A. De Luca, B. Bischl, H. Hussmann, M. Bühner, Personality traits predict smartphone usage. Eur. J. Pers. **31**(6), 701–722 (2017). https://doi.org/10.1002/per.2113

78. J. Teng, B. Zhang, X. Li, X. Bai, D. Xuan, E-shadow: lubricating social interaction using mobile phones, in *2011 31st International Conference on Distributed Computing Systems* (2011), pp. 909–918. https://doi.org/10.1109/ICDCS.2011.48

79. J. Teng, B. Zhang, X. Li, X. Bai, D. Xuan, E-shadow: lubricating social interaction using mobile phones. IEEE Trans. Comput. **63**(6), 1422–1433 (2014). https://doi.org/10.1109/TC.2012.290

80. Z. Yang, B. Zhang, J. Dai, A. Champion, D. Xuan, D. Li, E-SmallTalker: a distributed mobile system for social networking in physical proximity, in *2010 IEEE 30th International Conference on Distributed Computing Systems (ICDCS)* (IEEE, 2010), pp. 468–477. https://doi.org/10.1109/ICDCS.2010.56

81. T. Yarkoni, Psychoinformatics: new horizons at the interface of the psychological and computing sciences. Curr. Dir. Psychol. Sci. **21**(6), 391–397 (2012). https://doi.org/10.1177/0963721412457362

82. S. Yogesh, S. Abha, S. Priyanka, Mobile usage and sleep patterns among medical students. Indian J. Physiol. Pharmacol. **58**(1), 100–103 (2014)

83. X. Zhang, F. Zhuang, W. Li, H. Ying, H. Xiong, S. Lu, Inferring mood instability via smartphone sensing: a multi-view learning approach, in *Proceedings of the 27th ACM International Conference on Multimedia* (MM'19) (Association for Computing Machinery, 2019), pp. 1401–1409. https://doi.org/10.1145/3343031.3350957

84. T. Zhou, Y. Lu, The effects of personality traits on user acceptance of mobile commerce. Int. J. Hum.–Comput. Interact. **27**(6), 545–561 (2011). https://doi.org/10.1080/10447318.2011.555298

85. S. Zickau, F. Beierle, I. Denisow, Securing mobile cloud data with personalized attribute-based meta information, in *Proceedings of 3rd IEEE International Conference on Mobile Cloud Computing, Services, and Engineering (MobileCloud)* (IEEE, 2015), pp. 205–210. https://doi.org/10.1109/MobileCloud.2015.14

86. E. Zielinski, J. Schulz-Zander, M. Zimmermann, C. Schellenberger, A. Ramirez, F. Zeiger, M. Mormul, F. Hetzelt, F. Beierle, H. Klaus, H. Ruckstuhl, A. Artemenko, Secure real-time communication and computing infrastructure for industry 4.0—challenges and opportunities, in *Proceedings of 2019 International Conference on Networked Systems (NetSys)* (IEEE, 2019), pp. 1–6. https://doi.org/10.1109/NetSys.2019.8854499

87. J. Zimmermann, W.C. Woods, S. Ritter, M. Happel, O. Masuhr, U. Jaeger, C. Spitzer, A.G.C. Wright, Integrating structure and dynamics in personality assessment: first steps toward the development and validation of a personality dynamics diary. Psychol. Assess. **31**(4), 516–531 (2019). https://doi.org/10.1037/pas0000625

Part I
Mobile Sensing and Personality

Chapter 2
Overview

This part contributes to the scientific understanding of people and their behavior, specifically behavior in terms of smartphone usage. We address the research tasks related to research question A: the relationship between smartphone sensor and usage data and the user's personality. We give an overview about the related terms, showing the importance of the psychological concept of *personality*. We survey previous work and studies related to psychoinformatical research utilizing mobile crowdsensing in Chap. 3 (A-Task1). Based on our findings, we developed a mobile system that enables psychoinformatical research, TYDR (A-Task2). We present TYDR, as well as our privacy model PM-MoDaC for apps relating to the mobile collection of data (A-Task2), and evaluate our privacy model and what data users are willing to share with researchers in Chap. 4 (A-Task3). In Chap. 5, we analyze the relationship between personality traits and frequency and duration of smartphone usage based on data collected with TYDR (A-Task4).

© The Author(s), under exclusive license to Springer Nature Switzerland AG 2021
F. Beierle, *Integrating Psychoinformatics with Ubiquitous Social Networking*,
T-Labs Series in Telecommunication Services,
https://doi.org/10.1007/978-3-030-68840-0_2

Chapter 3
Related Work

First, we investigate related psychological research about personality (Sect. 3.1). Second, we give a brief summary about the fundamentals of context data (Sect. 3.2). Third, we give an extensive overview about existing research related to the collection of context data, focusing on psychoinformatical studies (Sect. 3.3). This section addresses research task A-Task1. Parts of this chapter have been published in [6, 7, 5].

3.1 Personality

Personality traits are patterns of thought, emotion, and behavior that are relatively consistent over time and across situations [27]. The most common model of individual differences in personality traits is the Big Five model, consisting of the traits *openness to experience*, *conscientiousness*, *extraversion*, *agreeableness*, and *neuroticism* [54, 42]. The five traits are empirically derived and reflect significant aspects in which individuals differ from each other.

People scoring high on the trait openness to experience exhibit the "ability and tendency to seek, detect, comprehend, utilize, and appreciate complex patterns of information, both sensory and abstract" [21]. Sometimes, this trait is referred to as *openness/intellect*. Conscientiousness is connected to self-discipline, industriousness, reliability, and a tendency to plan rather than be spontaneous [86]. Extraverts tend to be enthusiastic and to seek social interaction and external stimulus [54]. People who are agreeable tend to seek social harmony and are forgiving, kind, and considerate [86, 73]. Neuroticism is sometimes referred to as *emotional stability*. Neurotic people tend to experience more negative emotions such as anxiety, fear, frustration, depressive mood, or anger [86, 41]. There are several personality tests derived from this model, e.g., the Revised NEO Personality Inventory (NEO-PI-R

F. Beierle, *Integrating Psychoinformatics with Ubiquitous Social Networking*,
T-Labs Series in Telecommunication Services,
https://doi.org/10.1007/978-3-030-68840-0_3

[18]), the Big Five Inventory–2 (BFI-2 [80]), and the International Personality Item Pool (IPIP [28]).

The relevance for context-aware computing becomes apparent when looking into relationships between personality traits and everyday behavior and experiences [65]. There are vast bodies of research finding significant relationships.

This includes everyday preference for products or items like music, movies, clothing, or choice of newspapers. Preferred musical style is related to the extraversion trait, with extraverts preferring upbeat and energetic music [70, 71]. Karumur et al. show how movie rating patterns are significantly correlated with personality traits [44]. Extraverts and introverts tend to choose different styles of clothing [77]. Conscientious people are less likely to read tabloid newspapers [37].

Personality is also important for advertising and electronic or mobile commerce. Lu and Hsiao find that the extraversion scale influences the willingness to pay for social networking services and suggest different personality-trait-related strategies for monetization [51]. Regarding advertisements, Myers at al. report that customers react most positively when the ad matches their personality profile [64]. Similarly, the authors of [52] find that extraversion and openness to experience are relevant factors with respect to persuasion via online advertising. Zhou and Lu report that the adoption of mobile commerce is related to extraversion and neuroticism [97].

Another field that benefits from personality-trait-adapted approaches is the treatment of patients and, more generally, users of mHealth (mobile health) apps. Halko and Kientz found significant relationships between personality and the way persuasive mHealth apps are designed [30]. Roche et al. report that psychological assessment benefits from person-specific approaches, taking into account the patient's personality [57].

On a broader scale, personality traits are also correlated with political opinion, specifically openness and conscientiousness [53, 43]. Furthermore, conscientiousness predicts overall academic and workplace performance [36]. In Part II of this thesis, we will go into detail about research related to personality and interpersonal relationships, relevant for social networking scenarios.

Overall, the existing research demonstrates that a vast field of context-aware applications in multiple domains could benefit from knowing the user's personality. This includes recommender systems (e.g., for music, movies), ads and mobile commerce, and mHealth.

3.2   Context Data

Context is something that characterizes an entity that is relevant to the user or application [20]. Context thus is application-dependent: for an indoor mobile tour guide, for example, weather should not be considered as context [20].

Yurur et al. categorize context data sources into three types [94]:

- *Physical sensors* capture physical data, e.g., GPS for location or accelerometer for activity.
- *Virtual sensors* from software applications and/or services, e.g., manually set location or computation power of the device.
- *Logical sensors* are a combination of physical and virtual sensors with additional information through various sources like databases or log files.

The authors divide context into four categories:

- *Device Context*: net connectivity, communication cost
- *User Context*: profile, geographic position, neighbors, social situation
- *Physical Context*: temperature, noise level, light intensity, traffic conditions
- *Temporal Context*: day, week, month, season, year

However, this does not reflect the users' interaction with the smartphone. For example, the number of pictures taken or which apps a user is using may yield important information about his/her context. A user taking many pictures and using map applications might indicate that he/she is at an unfamiliar place that he/she enjoys.

3.3 Mobile Crowdsensing in Psychoinformatics

Psychoinformatics and digital phenotyping are interested in objective measurements of study participants' behaviors and surroundings. The smartphone and the context data it can sense are the optimal device to capture those. As introduced in Chap. 1, the collection of such context data is often referred to as mobile (crowd)sensing. There are two extremes on a scale of how such sensing can be designed. *Participatory sensing* focuses on the active collection and sharing of data by users [11]. Here, the user actively decides about the sensing process. *Opportunistic sensing* on the other hand refers to passive sensing with minimal user involvement [46]. Mobile crowdsensing for psychoinformatical studies typically falls in the category of opportunistic sensing, as the smartphone data correlated with psychological aspects about the user is sensed passively in the background.

Generally speaking, the more diverse the sensors and data types collected, the more precise the context of the user can be described. Context data can convey more information than their original purpose intended to. The location of a user might not only be useful for a navigational app but also convey information about his/her preferences. As another example, app usage statistics will give insight into the user's structure of the day. In this section, we give an overview about related work that previously dealt with the relationship between smartphone data and psychological aspects about the user. Some work from psychology/psychoinformatics focuses conceptually on what type of studies can be conducted and what open challenges there are from a psychology perspective [58, 33, 76, 72]. We want to focus here on concrete studies and investigate what context data points were actually recorded and

Table 3.1 Overview of data sources and static user information that were correlated in previous studies

Data sources	User information	Ref.	Year	n
Application usage (pre-smartphone), Bluetooth, calling profiles, calls, SMS	Personality traits	[15]	2011	83
Application usage (pre-smartphone), Bluetooth, calling profiles, calls, SMS	Personality traits	[16]	2013	117
Calls, location, SMS	Personality traits	[19]	2013	69
Calls, SMS	Personality traits	[62]	2014	49
WhatsApp usage	Age, gender, education, personality traits	[63]	2015	2418
Location	Personality traits	[17]	2015	174
Display state, location	Depression	[74]	2015	28
Installed apps	Personality traits	[93]	2016	2043
Location	Social anxiety	[38]	2016	18
App usage, Bluetooth, calls, location	Sociability	[22]	2016	10
Calls, location, SMS	Cooperation attitude	[78]	2016	54
App usage, notification metadata	Depression	[55]	2016	25
Technology usage times	Personality traits	[29]	2017	62
App usage	Personality traits	[82]	2017	137
Calls, SMS	Social capital	[79]	2017	55
Calls, location, SMS	Social anxiety	[10]	2017	54
Location	Personality traits	[45]	2018	20
Bluetooth, calls, location, SMS	Personality traits	[60]	2018	636
Location, microphone (ambient sound, ambient voice), phone usage, physical activity	Personality traits	[90]	2018	159
App usage, calls, location	Sensation seeking	[75]	2018	260
App usage, battery status, Bluetooth, calls, display state, location, photo metadata, SMS, WiFi	Personality traits	[81]	2019	624
Calls, media app usage, messaging app usage, SMS	Sociability	[31]	2019	926[a]
Calls	Personality traits	[61]	2019	106

[a]Combined dataset from four studies

set into relation with psychological aspects. Additionally, we survey other related works dealing with the collection of mobile data.

In Table 3.1, we give an overview of related studies that correlated sensor and/or smartphone usage data with user information related to personality or other psychological concepts. In this table, we look at studies that dealt with rather static aspects of participants, i.e., aspects that are considered rather stable and that do not change over time. Note that the first two studies have been conducted with feature phones, before the advent of smartphones [15, 16]. There are some additional studies correlating personality traits and phone usage that did not collect data from feature or smartphones but relied on self-reports of users [66, 12, 47].

The *data sources* given in the table differ in their level. For example, accelerometer data is low-level sensor data, while the current activity (e.g., walking, in car, etc.) or a daily step count is higher-level sensor data that utilizes accelerometer data. The available data sources depend on the used mobile OS and on the available libraries and software development kits (SDKs). In the table, we list the sources mentioned in the cited papers. There might be some steps in between low-level sensor data and the user's personality, like estimating the user's sleep pattern from low-level sensor data like phone lock/unlock events. For overviews related to determining higher-level features from lower-level sensor data, see, e.g., [33, 32, 59]. While some of the studies are only interested in specific data types, others track a wider range of data sources.

The ground truth for *user information* is typically assessed via self-report methods, i.e., questionnaires. Most of the studies are done in the intersection of computer science and psychological research with objective behavioral measurements (e.g., [62, 38, 55]). Others deal with context-aware recommender systems, e.g., recommending new apps based on the personality correlated with already installed apps [93]. Some studies go further than correlating automatically collected data with the personality of the user.[1] The StudentLife project, for example, collected sensor data and queried student participants with a variety of questionnaires to predict mental health and academic performance [88, 87, 89].

We sorted the entries in Table 3.1 by the year the corresponding paper was released. The column n refers to the sample size reported in the paper. We list the sample size that was reported to be used in the data analysis, which might be smaller than the number of initially acquired users/participants. For more than half of the listed studies, in 12 out of 23 papers, the sample size $n < 100$ (range 10–83). The median n of the listed studies is 83. In general, the higher the n, the higher the validity of the findings. We observe two outliers with exceptionally high sample sizes: [63] with 2418 users and [93] with 2043 users. Installed apps (cf. [93]) could be instantly recorded once. Most likely, this made it easier to gather than data sources that require longer processes or have to be recorded for a longer time period. One reason for rather low values of n could be connected to privacy concerns of the users. Some data might be more difficult to collect than others. Note, for example, that only one of the studies used the microphone [90]. Based on a sample size of $n = 1560$, in Sect. 4.7, we will analyze, based on our own mobile context data collection app TYDR, what users are willing to share which kind of data.

With 14 out of 23, about 60% of the listed studies are related to personality traits of the user. Other topics that are covered in more than one paper are depression, social anxiety, and sociability (two papers each, cf. Table 3.1). Some studies on pre-smartphone-era cell phones found correlations between personality traits and mobile phone usage. For example, these are the results of a study done on the general use of mobile phones (calls, text messages, changing ringtones and wallpapers) [12], as well as of a study about using mobile phone games [66]. While these studies

---

[1] We did not list those studies in the table.

were based on self-reports by users, Chittaranjan et al. conducted two user studies
in which they collected usage data on Nokia N95 phones [15, 16]. In those studies,
the authors were looking at Bluetooth scan data, call logs, text messages, calling
profiles, and application usage. At the time the study was conducted, apps were not
as common as nowadays with Android and iOS. The authors state that "features
derived from the App Logs were sparse due to the low frequency of usage of some
of the applications" [16]. Since the publication of that study (2013), app usage is
much more commonplace. In more recent studies, relations between smartphone
usage and the five personality dimensions were shown [82, 90, 81]. One of the most
recent studies in the table, published in 2019, after our development of TYDR was
finished, contains the most variety of data sources (see [81] in Table 3.1).

In Table 3.2, we list related studies that deal with collecting smartphone data and
finding correlations with rather dynamic psychological aspects about the user that
change over time. Such change can be in rather short time frames (e.g., emotions) to
hours, daily changes, or 14-day intervals (some of depression-related studies, e.g.,

Table 3.2 Overview of data sources and dynamic user information that were correlated in previous studies

Data sources	User information	Ref.	Year	n
Accelerometer, Bluetooth, location	Emotions	[69]	2010	12
App usage, calls, email, location, SMS, websites	Mood	[50]	2013	32
Bluetooth, calls, SMS, weather	Daily stress	[9]	2014	117
Location	Depressive states	[13]	2015	28
Calls, location, SMS	Stress, depression, loneliness	[8]	2015	47
App usage, calls, display state, light sensor, microphone, network traffic, physical activity, SMS	Stress	[84]	2015	15
Accelerometer, app usage, calls, display state, SMS	Mood	[4]	2016	27
Accelerometer, app usage, calls, display state, photo metadata, SMS	Mood	[3]	2016	27
Location, physical activity	Depression	[23]	2016	79
App usage, calls, display state	Depression, anxiety, stress	[39]	2016	18
Accelerometer, keypress entry time	Depression	[14]	2017	20
App usage, Bluetooth, calls, location, physical activity, SMS, WiFi	Emotions	[85]	2017	10
Location, notification metadata, physical activity	Mood	[56]	2017	28
Accelerometer, app usage, calls, compass, display sate, gyroscope, light sensor, location, microphone, WiFi, SMS	Compound emotion	[96]	2018	30
Accelerometer, app usage, display state, light sensor, microphone, WiFi	Mood instability	[95]	2019	68
Location	Stress	[67]	2019	77

[23]). With respect to the performed data analysis, this means that the correlations to be looked for are not based on the overall phone usage but on the specific intervals. The psychological aspects that researchers are interested in cover a range of different topics. The most common ones are mood, depression, and stress, which are the focus topics in about a third of the related studies (5/16). Correlations between emotions and smartphone usage are researched in 3/16 studies. To assess most of these aspects, psychometric questionnaires are utilized to label data. The related research method is sometimes referred to as *ecological momentary assessment* (EMA) [83]. The idea is that participants repeatedly report their experiences in real-world settings, avoiding retrospective bias [68]. Another related research method is the *experience sampling method* (ESM) [49]. The idea is also to assess momentary behavior, feelings, or symptoms, typically via self-reports.

With respect to the collection of data, this means that users have to spend more effort in manually annotating the collected smartphone context data. This immediately explains the lower sample sizes. Users might not sign up for such studies, drop out, or just stop answering the questionnaires. For all except one of the studies (15/16 papers), the sample size $n < 100$. The median sample size n is 28. Again, there is also a dependency between recruitment of users/study participants and the type of data that is being collected. For example, the collection of (meta)data about keypresses (cf. [14] in Table 3.2) requires a custom keyboard software, the installation of which requires high degrees of trust because such software can log any keypress. Additionally, regarding the data sources, we observe that for the studies listed in Table 3.2, there are more sources that can track changes in high frequencies, e.g., accelerometer, gyroscope, light sensor, or microphone.

To recognize how sensitive the collected data might be, consider, for example, that when combining data from multiple sources, researchers can see for a user which app he/she used for how long at which location and what brightness the location had. As we are dealing with such highly sensitive data, privacy concerns should have a high priority. Merriam-Webster defines *privacy* as "the state of being apart from company or observation" and "freedom from unauthorized intrusion."[2] This can seem quite fuzzy when applied in the context of mobile data collection. Regional data protection regulations like the European Union's GDPR (General Data Protection Regulation) try to make privacy measures more concrete. However, uncertainties remain regarding the specifics on how to comply to those regulations [25]. Only 2 out of the 39 papers (5%) reviewed in Tables 3.1 and 3.2 give somewhat detailed information about privacy awareness [61, 31]. Both of these papers were published after we published our privacy model in [6]. Harari et al. [31] actually use our privacy model to describe the privacy mechanism they implemented. Furthermore, 9 out of 39 papers (23%) give no technical details about privacy awareness [9, 45, 4, 29, 95, 10, 19, 55, 69] and in the case of [69] even explicitly disregard privacy: "privacy is not a major concern for this system, since all users voluntarily agree to carry the devices for constant monitoring." The rest of the

---

[2]http://www.merriam-webster.com/dictionary/privacy.

papers, 28 out of 39 papers (72%), give some details, mostly mentioning one or two aspects like data anonymization or approval by an ethics commission/institutional review board. In their study about technologies related to self-reporting of emotions, Fuentes et al. report similar findings [26]: most of the reviewed work did not mention privacy at all. In their review of mobile sensing systems, Laport-López et al. find that only 13% of the reviewed works implement privacy measures [48].

From a commercial side, there are several related apps available, for example, Moodpath[3] or Daylio[4] for tracking mood. Both use a freemium business model, and both focus on tracking the user's mood (and daily activities and experiences in the case of Daylio). These apps, however, do not follow the idea of collecting smartphone context data.

Besides the apps used in the studies listed above, there are further apps and software frameworks and libraries related to the collection of mobile context data. When focusing on the benefit for the user, sometimes the term *quantified self* is used to refer to those kinds of apps, highlighting the quantification of aspects of daily life.

The Menthal app attracted the highest amount of users we found in academic literature of related apps.[5] The authors report about more than 400,000 app installations [2]. In [1], the authors analyze the relationship between age, gender, and app usage with a sample size of 30,677 users, without going into detail regarding psychological aspects. In another study, they show how almost all of their users are individually identifiable by the combination of apps they use [91]. Menthal is one of the few apps we found in the related work that was available in the app store for the general public while also having publications about them. A paper published in 2019 (after the publication of TYDR) mentions an app called Insights[6] that tracks a variety of data sources (contact list, calls, SMS, display state, battery state, installed apps, app usage, location, and data traffic) [61].

Some researchers developed frameworks or libraries that allow other researchers to conduct studies related to the collection of mobile context data. Sensus [92], LiveLabs [40], and AWARE [24] are some examples. As far as the papers and website indicate, none of these frameworks provide support for collecting music and photo metadata, which we enable with TYDR.

In [35], the authors present a framework called *BaranC* for monitoring and analyzing interactions of users with their smartphones. The goal of BaranC is to offer personalized services. In another work by the same authors, an application utilizing their framework is presented [34], predicting the next application a user will use.

---

[3]https://mymoodpath.com/.

[4]https://daylio.webflow.io/.

[5]https://menthal.org/.

[6]Android app; more info at https://www.insightsapp.org/.

References

1. I. Andone, K. Błaszkiewicz, M. Eibes, B. Trendafilov, C. Montag, A. Markowetz, How age and gender affect smartphone usage, in *Proceedings of the 2016 ACM International Joint Conference on Pervasive and Ubiquitous Computing: Adjunct.* UbiComp'16 (ACM, 2016), pp. 9–12. https://doi.org/10.1145/2968219.2971451

2. I. Andone, K. Blaszkiewicz, M. Eibes, B. Trendafilov, C. Montag, A. Markowetz, Menthal: quantifying smartphone usage, in *Proceedings of the 2016 ACM International Joint Conference on Pervasive and Ubiquitous Computing: Adjunct.* UbiComp'16 (ACM, 2016), pp. 559–564. https://doi.org/10.1145/2968219.2968321

3. J. Asselbergs, J. Ruwaard, M. Ejdys, N. Schrader, M. Sijbrandij, H. Riper, Mobile phone-based unobtrusive ecological momentary assessment of day-to-day mood: an explorative study. J. Med. Internet Res. **18**(3), e72 (2016). https://doi.org/10.2196/jmir.5505

4. D. Becker, V. Bremer, B. Funk, J. Asselbergs, H. Riper, J. Ruwaard, How to predict mood? Delving into features of smartphone-based data, in *Twenty-Second Americas Conference on Information Systems (AMCIS)* (2016), pp. 1–10

5. F. Beierle, S.C. Matz, M. Allemand, Smartphone sensing in personality science, in *Mobile Sensing in Psychology: Methods and Applications*, ed. by M.R. Mehl, C. Wrzus, M. Eid, G. Harari, U.E. Priemer (Guilford Press, New York City, 2021)

6. F. Beierle, V.T. Tran, M. Allemand, P. Neff, W. Schlee, T. Probst, R. Pryss, J. Zimmermann, Context data categories and privacy model for mobile data collection apps. Proc. Comput. Sci. The 15th International Conference on Mobile Systems and Pervasive Computing (MobiSPC) **134**, 18–25 (2018). https://doi.org/10.1016/j.procs.2018.07.139

7. F. Beierle, V.T. Tran, M. Allemand, P. Neff, W. Schlee, T. Probst, J. Zimmermann, R. Pryss, What data are smartphone users willing to share with researchers? J. Ambient Intell. Humaniz. Comput. **11**, 2277–2289 (2020). https://doi.org/10.1007/s12652-019-01355-6

8. D. Ben-Zeev, E.A. Scherer, R. Wang, H. Xie, A.T. Campbell, Next-generation psychiatric assessment: using smartphone sensors to monitor behavior and mental health. Psychiatr. Rehabil. J. **38**(3), 218–226 (2015). https://doi.org/10.1037/prj0000130

9. A. Bogomolov, B. Lepri, M. Ferron, F. Pianesi, A.S. Pentland, Daily stress recognition from mobile phone data, weather conditions and individual traits, in *Proceedings of the 22nd ACM International Conference on Multimedia.* MM'14 (ACM, 2014), pp. 477–486. https://doi.org/10.1145/2647868.2654933

10. M. Boukhechba, Y. Huang, P. Chow, K. Fua, B.A. Teachman, L.E. Barnes, Monitoring social anxiety from mobility and communication patterns, in *Proceedings of the 2017 ACM International Joint Conference on Pervasive and Ubiquitous Computing and Proceedings of the 2017 ACM International Symposium on Wearable Computers.* UbiComp'17 (ACM, 2017), pp. 749–753. https://doi.org/10.1145/3123024.3125607

11. J. Burke, D. Estrin, M. Hansen, A. Parker, N. Ramanathan, S. Reddy, M.B. Srivastava, Participatory sensing, in *Workshop on World-Sensor-Web (WSW'06): Mobile Device Centric Sensor Networks and Applications* (ACM, 2006), pp. 117–134

12. S. Butt, J.G. Phillips, Personality and self reported mobile phone use. Comput. Hum. Behav. **24**(2), 346–360 (2008). Part Special Issue: Cognition and Exploratory Learning in Digital Age. https://doi.org/10.1016/j.chb.2007.01.019

13. L. Canzian, M. Musolesi, Trajectories of depression: unobtrusive monitoring of depressive states by means of smartphone mobility traces analysis, in *Proceedings of the 2015 ACM International Joint Conference on Pervasive and Ubiquitous Computing.* UbiComp'15 (ACM, 2015), pp. 1293–1304. https://doi.org/10.1145/2750858.2805845

14. B. Cao, L. Zheng, C. Zhang, P.S. Yu, A. Piscitello, J. Zulueta, O. Ajilore, K. Ryan, A.D. Leow, DeepMood: modeling mobile phone typing dynamics for mood detection, in *Proceedings of the 23rd ACM SIGKDD International Conference on Knowledge Discovery and Data Mining.* KDD'17 (Association for Computing Machinery, 2017), pp. 747–755. https://doi.org/10.1145/3097983.3098086

15. G. Chittaranjan, J. Blom, D. Gatica-Perez, Who's who with big-five: analyzing and classifying personality traits with smartphones, in *Proceedings of 2011 15th Annual International Symposium on Wearable Computers* (IEEE, 2011), pp. 29–36. https://doi.org/10.1109/ISWC. 2011.29
16. G. Chittaranjan, J. Blom, D. Gatica-Perez, Mining large-scale smartphone data for personality studies. Pers. Ubiquit. Comput. **17**(3), 433–450 (2013). https://doi.org/10.1007/s00779-011-0490-1
17. M.J. Chorley, R.M. Whitaker, S.M. Allen, Personality and location-based social networks. Comput. Hum. Behav. **46**(Supplement C), 45–56 (2015). https://doi.org/10.1016/j.chb.2014. 12.038
18. P.T. Costa Jr, R.R. McCrae, The revised NEO personality inventory (NEO-PI-R), in *The SAGE Handbook of Personality Theory and Assessment: Volume 2—Personality Measurement and Testing* (SAGE Publications Ltd, 2008), pp. 179–198. https://doi.org/10.4135/9781849200479
19. Y.-A. de Montjoye, J. Quoidbach, F. Robic, A.S. Pentland, Predicting personality using novel mobile phone-based metrics, in *Social Computing, Behavioral-Cultural Modeling and Prediction*, ed. by A.M. Greenberg, W.G. Kennedy, N.D. Bos. Lecture Notes in Computer Science (Springer, 2013), pp. 48–55. https://doi.org/10.1007/978-3-642-37210-0_6
20. A.K. Dey, Understanding and using context. Pers. Ubiquit. Comput. **5**(1), 4–7 (2001). https://doi.org/10.1007/s007790170019
21. C.G. DeYoung, Openness/intellect: a dimension of personality reflecting cognitive exploration, in *APA Handbook of Personality and Social Psychology, Volume 4: Personality Processes and Individual Differences*, ed. by M. Mikulincer, P.R. Shaver, M.L. Cooper, R.J. Larsen (American Psychological Association, 2015), pp. 369–399. https://doi.org/10.1037/14343-017
22. P. Eskes, M. Spruit, S. Brinkkemper, J. Vorstman, M.J. Kas, The sociability score: app-based social profiling from a healthcare perspective. Comput. Hum. Behav. **59**, 39–48 (2016). https://doi.org/10.1016/j.chb.2016.01.024
23. A.A. Farhan, C. Yue, R. Morillo, S. Ware, J. Lu, J. Bi, J. Kamath, A. Russell, A. Bamis, B. Wang, Behavior vs. introspection: refining prediction of clinical depression via smartphone sensing data, in *2016 IEEE Wireless Health (WH)* (IEEE, 2016), pp. 1–8. https://doi.org/10. 1109/WH.2016.7764553
24. D. Ferreira, V. Kostakos, A.K. Dey, AWARE: mobile context instrumentation framework, in *Frontiers in ICT*, vol. 2 (2015). https://doi.org/10.3389/fict.2015.00006
25. A.R. Filippo, Innovating in uncertainty: effective compliance and the GDPR. Harv. J. Law Technol. (2018). https://jolt.law.harvard.edu/digest/innovating-in-uncertainty-effective-compliance-and-the-gdpr
26. C. Fuentes, V. Herskovic, I. Rodríguez, C. Gerea, M. Marques, P.O. Rossel, A systematic literature review about technologies for self-reporting emotional information. J. Ambient Intell. Humaniz. Comput. **8**(4), 593–606 (2017). https://doi.org/10.1007/s12652-016-0430-z
27. D.C. Funder, Accurate personality judgment. Curr. Dir. Psychol. Sci. **21**(3), 177–182 (2012). https://doi.org/10.1177/0963721412445309
28. L.R. Goldberg, J.A. Johnson, H.W. Eber, R. Hogan, M.C. Ashton, C.R. Cloninger, H.G. Gough, The international personality item pool and the future of public-domain personality measures. J. Res. Pers. **40**(1), 84–96 (2006). Proceedings of the 2005 Meeting of the Association of Research in Personality. https://doi.org/10.1016/j.jrp.2005.08.007
29. T. Grover, G. Mark, Digital footprints: predicting personality from temporal patterns of technology use, in *Proceedings of the 2017 ACM International Joint Conference on Pervasive and Ubiquitous Computing and Proceedings of the 2017 ACM International Symposium on Wearable Computers*. UbiComp'17 (ACM, 2017), pp. 41–44. https://doi.org/10.1145/3123024. 3123139
30. S. Halko, J.A. Kientz, Personality and persuasive technology: an exploratory study on health-promoting mobile applications, in *Persuasive Technology*, ed. by D. Hutchison, T. Kanade, J. Kittler, J.M. Kleinberg, F. Mattern, J.C. Mitchell, M. Naor, O. Nierstrasz, C. Pandu Rangan, B. Steffen, M. Sudan, D. Terzopoulos, D. Tygar, M.Y. Vardi, G. Weikum, T. Ploug, P. Hasle, H. Oinas-Kukkonen, vol. 6137 (Springer, Berlin/Heidelberg, 2010), pp. 150–161. https://doi.org/10.1007/978-3-642-13226-1_16

31. G.M. Harari, S.R. Müller, C. Stachl, R. Wang, W. Wang, M. Bühner, P.J. Rentfrow, A.T. Campbell, S.D. Gosling, Sensing sociability: individual differences in young adults' conversation, calling, texting, and app use behaviors in daily life. J. Pers. Soc. Psychol. (2019). https://doi.org/10.1037/pspp0000245

32. G.M. Harari, S.R. Müller, M.S. Aung, P.J. Rentfrow, Smartphone sensing methods for studying behavior in everyday life. Curr. Opin. Behav. Sci. **18**(Supplement C), 83–90 (2017). SI: 18: Big Data in the Behavioural Sciences (2017). https://doi.org/10.1016/j.cobeha.2017.07.018.

33. G.M. Harari, N.D. Lane, R. Wang, B.S. Crosier, A.T. Campbell, S.D. Gosling, Using smartphones to collect behavioral data in psychological science: opportunities, practical considerations, and challenges. Perspect. Psychol. Sci. **11**(6), 838–854 (2016). https://doi.org/10.1177/1745691616650285

34. M. Hashemi, J. Herbert, A next application prediction service using the BaranC framework, in *Proceedings of 2016 IEEE 40th Annual Computer Software and Applications Conference (COMPSAC)* (IEEE, 2016), pp. 519–523. https://doi.org/10.1109/COMPSAC.2016.30

35. M. Hashemi, J. Herbert, User interaction monitoring and analysis framework, in *Proceedings of 2016 IEEE/ACM International Conference on Mobile Software Engineering and Systems (MobileSoft'16)* (ACM, 2016), pp. 7–8. https://doi.org/10.1145/2897073.2897108

36. D.M. Higgins, J.B. Peterson, R.O. Pihl, A.G.M. Lee, Prefrontal cognitive ability intelligence, big five personality and the prediction of advanced academic and workplace performance. J. Pers. Soc. Psychol. **93**(2), 298–319 (2007). https://doi.org/10.1037/0022-3514.93.2.298

37. J.B. Hirsh, C.G. DeYoung, J.B. Peterson, Metatraits of the big five differentially predict engagement and restraint of behavior. J. Pers. **77**(4), 1085–1102 (2009). https://doi.org/10.1111/j.1467-6494.2009.00575.x

38. Y. Huang, H. Xiong, K. Leach, Y. Zhang, P. Chow, K. Fua, B.A. Teachman, L.E. Barnes, Assessing social anxiety using GPS trajectories and point-Of-interest data, in *Proceedings of the 2016 ACM International Joint Conference on Pervasive and Ubiquitous Computing*. UbiComp'16 (Association for Computing Machinery, 2016), pp. 898–903. https://doi.org/10.1145/2971648.2971761

39. G.C.-L. Hung, P.-C. Yang, C.-C. Chang, J.-H. Chiang, Y.-Y. Chen, Predicting negative emotions based on mobile phone usage patterns: an exploratory study. JMIR Res. Protoc. **5**(3), e160 (2016). https://doi.org/10.2196/resprot.5551

40. K. Jayarajah, R.K. Balan, M. Radhakrishnan, A. Misra, Y. Lee, LiveLabs: building in-situ mobile sensing & behavioural experimentation TestBeds, in *Proceedings of the 14th Annual International Conference on Mobile Systems, Applications, and Services*. MobiSys'16 (ACM, 2016), pp. 1–15. https://doi.org/10.1145/2906388.2906400

41. B.F. Jeronimus, H. Riese, R. Sanderman, J. Ormel, Mutual reinforcement between neuroticism and life experiences: a five-wave, 16-year study to test reciprocal causation. J. Pers. Soc. Psychol. **107**(4), 751–764 (2014). https://doi.org/10.1037/a0037009

42. O.P. John, S. Srivastava, The big five trait taxonomy: history measurement, and theoretical perspectives, in *Handbook of Personality: Theory and Research*, 2nd edn. (Guilford Press, 1999), pp. 102–138

43. J.T. Jost, The end of the end of ideology. Am. Psychol. **61**(7), 651–670 (2006). https://doi.org/10.1037/0003-066X.61.7.651

44. R.P. Karumur, T.T. Nguyen, J.A. Konstan, Personality user preferences and behavior in recommender systems. Inf. Syst. Front. 1–25 (2017). https://doi.org/10.1007/s10796-017-9800-0

45. S.Y. Kim, H.J. Koo, H.Y. Song, A study on estimation of human personality from location visiting preference. J. Ambient Intell. Humaniz. Comput. **9**(3), 629–642 (2018). https://doi.org/10.1007/s12652-017-0459-7

46. N.D. Lane, E. Miluzzo, H. Lu, D. Peebles, T. Choudhury, A.T. Campbell, A survey of mobile phone sensing. IEEE Commun. Mag. **48**(9), 140–150 (2010). https://doi.org/10.1109/MCOM.2010.5560598

47. W. Lane, C. Manner, The influence of personality traits on mobile phone application preferences. J. Econ. Behav. Stud. **4**(5), 252–260 (2012)

48. F. Laport-López, E. Serrano, J. Bajo, A.T. Campbell, A review of mobile sensing systems, applications, and opportunities. Knowl. Inf. Syst. (2019). https://doi.org/10.1007/s10115-019-01346-1
49. R. Larson, M. Csikszentmihalyi, The experience sampling method, in *Flow and the Foundations of Positive Psychology: The Collected Works of Mihaly Csikszentmihalyi*, ed. by H. Reis. Flow and the Foundations of Positive Psychology (Springer, 2014), pp. 41–56. https://doi.org/10.1007/978-94-017-9088-8_2
50. R. LiKamWa, Y. Liu, N.D. Lane, L. Zhong, MoodScope: building a mood sensor from smartphone usage patterns, in *Proceeding of the 11th Annual International Conference on Mobile Systems, Applications, and Services*. MobiSys'13 (ACM, 2013), pp. 389–402. https://doi.org/10.1145/2462456.2464449
51. H.-P. Lu, K.-L. Hsiao, The influence of extro/introversion on the intention to pay for social networking sites. Inf. Manag. **47**(3), 150–157 (2010). https://doi.org/10.1016/j.im.2010.01.003
52. S.C. Matz, M. Kosinski, G. Nave, D.J. Stillwell, Psychological targeting as an effective approach to digital mass persuasion. Proc. Natl. Acad. Sci. **114**(48), 12714–12719 (2017). https://doi.org/10.1073/pnas.1710966114
53. R.R. McCrae, Social consequences of experiential openness. Psychol. Bull. **120**(3), 323–337 (1996). https://doi.org/10.1037/0033-2909.120.3.323
54. R.R. McCrae, O.P. John, An introduction to the five-factor model and its applications. J. Pers. **60**(2), 175–215 (1992)
55. A. Mehrotra, R. Hendley, M. Musolesi, Towards multi-modal anticipatory monitoring of depressive states through the analysis of human-smartphone interaction, in *Proceedings of the 2016 ACM International Joint Conference on Pervasive and Ubiquitous Computing: Adjunct*. UbiComp'16 (Association for Computing Machinery, 2016), pp. 1132–1138. https://doi.org/10.1145/2968219.2968299
56. A. Mehrotra, F. Tsapeli, R. Hendley, M. Musolesi, MyTraces: investigating correlation and causation between users' emotional states and mobile phone interaction. Proc. ACM Interact. Mob. Wearable Ubiquit. Technol. **1**(3), 83:1–83:21 (2017). https://doi.org/10.1145/3130948
57. M.J. Roche, A.L. Pincus, A.L. Rebar, D.E. Conroy, N. Ram, Enriching psychological assessment using a person-specific analysis of interpersonal processes in daily life. Assessment **21**(5), 515–528 (2014). https://doi.org/10.1177/1073191114540320
58. G. Miller, The smartphone psychology manifesto. Perspect. Psychol. Sci. **7**(3), 221–237 (2012). https://doi.org/10.1177/1745691612441215
59. D.C. Mohr, M. Zhang, S.M. Schueller, Personal sensing: understanding mental health using ubiquitous sensors and machine learning. Annu. Rev. Clin. Psychol. **13**, 23–47 (2017). https://doi.org/10.1146/annurev-clinpsy-032816-044949
60. B. Mønsted, A. Mollgaard, J. Mathiesen, Phone-based metric as a predictor for basic personality traits. J. Res. Pers. **74**, 16–22 (2018). https://doi.org/10.1016/j.jrp.2017.12.004
61. C. Montag, H. Baumeister, C. Kannen, R. Sariyska, E.-M. Meßner, M. Brand, Concept, possibilities and pilot-testing of a new smartphone application for the social and life sciences to study human behavior including validation data from personality psychology. J. Multidiscip. Sci. J. **2**(2), 102–115 (2019). https://doi.org/10.3390/j2020008
62. C. Montag, K. Błaszkiewicz, B. Lachmann, I. Andone, R. Sariyska, B. Trendafilov, M. Reuter, A. Markowetz, Correlating personality and actual phone usage. J. Individ. Differ. **35**(3), 158–165 (2014). https://doi.org/10.1027/1614-0001/a000139
63. C. Montag, K. Błaszkiewicz, R. Sariyska, B. Lachmann, I. Andone, B. Trendafilov, M. Eibes, A. Markowetz, Smartphone usage in the 21st century: who is active on WhatsApp? BMC Res. Notes **8**(1), 331 (2015). https://doi.org/10.1186/s13104-015-1280-z
64. S.D. Myers, S. Sen, A. Alexandrov, The moderating effect of personality traits on attitudes toward advertisements: a contingency framework. Manag. Mark. **5**(3), 3–20 (2010)
65. D.J. Ozer, V. Benet-Martínez, Personality and the prediction of consequential outcomes. Annu. Rev. Psychol. **57**(1), 401–421 (2005). https://doi.org/10.1146/annurev.psych.57.102904.190127
66. J.G. Phillips, S. Butt, A. Blaszczynski, Personality and self-reported use of mobile phones for games. CyberPsychol. Behav. **9**(6), 753–758 (2006). https://doi.org/10.1089/cpb.2006.9.753

67. R. Pryss, D. John, M. Reichert, B. Hoppenstedt, L. Schmid, W. Schlee, M. Spiliopoulou, J. Schobel, R. Kraft, M. Schickler, B. Langguth, T. Probst, Machine learning findings on geospatial data of users from the TrackYourStress mHealth Crowdsensing Platform, in *2019 IEEE 20th International Conference on Information Reuse and Integration for Data Science (IRI)* (2019), pp. 350–355. https://doi.org/10.1109/IRI.2019.00061

68. R. Pryss, T. Probst, W. Schlee, J. Schobel, B. Langguth, P. Neff, M. Spiliopoulou, M. Reichert, Prospective crowdsensing versus retrospective ratings of tinnitus variability and tinnitus–stress associations based on the TrackYourTinnitus Mobile Platform. Int. J. Data Sci. Anal. **8**(4), 327–338 (2019). https://doi.org/10.1007/s41060-018-0111-4

69. K.K. Rachuri, M. Musolesi, C. Mascolo, P.J. Rentfrow, C. Longworth, A. Aucinas, EmotionSense: a mobile phones based adaptive platform for experimental social psychology research, in *Proceedings of the 12th ACM International Conference on Ubiquitous Computing*. UbiComp'10 (ACM, 2010), pp. 281–290. https://doi.org/10.1145/1864349.1864393

70. D. Rawlings, V. Ciancarelli, Music preference and the five-factor model of the NEO personality inventory. Psychol. Music **25**(2), 120–132 (1997). https://doi.org/10.1177/0305735697252003

71. P.J. Rentfrow, S.D. Gosling, The Do Re Mi's of everyday life: the structure and personality correlates of music preferences. J. Pers. Soc. Psychol. **84**(6), 1236–1256 (2003). https://doi.org/10.1037/0022-3514.84.6.1236

72. J. Rooksby, A. Morrison, D. Murray-Rust, Student perspectives on digital phenotyping: the acceptability of using smartphone data to assess mental health, in *Proceedings of the 2019 CHI Conference on Human Factors in Computing Systems*. CHI'19 (ACM, 2019), pp. 425:1–425:14. https://doi.org/10.1145/3290605.3300655

73. S. Rothmann, E.P. Coetzer, The big five personality dimensions and job performance. SA J. Ind. Psychol. **29**(1) (2003). https://doi.org/10.4102/sajip.v29i1.88

74. S. Saeb, M. Zhang, C.J. Karr, S.M. Schueller, M.E. Corden, K.P. Kording, D.C. Mohr, Mobile phone sensor correlates of depressive symptom severity in daily-life behavior: an exploratory study. J. Med. Internet Res. **17**(7), e175 (2015). https://doi.org/10.2196/jmir.4273

75. R. Schoedel, Q. Au, S.T. Völkel, F. Lehmann, D. Becker, M. Bühner, B. Bischl, H. Hussmann, C. Stachl, Digital footprints of sensation seeking. Zeitschrift für Psychologie **226**(4), 232–245 (2018). https://doi.org/10.1027/2151-2604/a000342

76. A. Seifert, M. Hofer, M. Allemand, Mobile data collection: smart, but not (yet) smart enough. Front. Neurosci. **12** (2018). https://doi.org/10.3389/fnins.2018.00971

77. R.S. Sharma, Clothing behaviour, personality and values: a correlational study. Psychol. Stud. **25**(2), 137–142 (1980)

78. V.K. Singh, R.R. Agarwal, Cooperative phoneotypes: exploring phone-based behavioral markers of cooperation, in *Proceedings of the 2016 ACM International Joint Conference on Pervasive and Ubiquitous Computing*. UbiComp'16 (ACM, 2016), pp. 646–657. https://doi.org/10.1145/2971648.2971755

79. V.K. Singh, I. Ghosh, Inferring individual social capital automatically via phone logs. Proc. ACM Hum.-Comput. Interact. **1**(CSCW), 95:1–95:12 (2017). https://doi.org/10.1145/3134730

80. C.J. Soto, O.P. John, The next big five inventory (BFI-2): developing and assessing a hierarchical model with 15 facets to enhance bandwidth, fidelity and predictive power. J. Pers. Soc. Psychol. **113**(1), 117–143 (2017). https://doi.org/10.1037/pspp0000096

81. C. Stachl, Q. Au, R. Schoedel, D. Buschek, S. Völkel, T. Schuwerk, M. Oldemeier, T. Ullmann, H. Hussmann, B. Bischl, M. Bühner, *Behavioral Patterns in Smartphone Usage Predict Big Five Personality Traits*. Preprint. PsyArXiv, (2019). https://doi.org/10.31234/osf.io/ks4vd

82. C. Stachl, S. Hilbert, J.-Q. Au, D. Buschek, A. De Luca, B. Bischl, H. Hussmann, M. Bühner, Personality traits predict smartphone usage. Eur. J. Pers. **31**(6), 701–722 (2017). https://doi.org/10.1002/per.2113

83. A.A. Stone, S. Shiffman, Ecological momentary assessment (EMA) in behavioral medicine. Ann. Behav. Med. **16**(3), 199–202 (1994). https://doi.org/10.1093/abm/16.3.199

84. T. Stütz, T. Kowar, M. Kager, M. Tiefengrabner, M. Stuppner, J. Blechert, F.H. Wilhelm, S. Ginzinger, Smartphone based stress prediction, in *User Modeling, Adaptation and Personalization*, ed. by F. Ricci, K. Bontcheva, O. Conlan, S. Lawless. Lecture Notes in Computer

Science (Springer International Publishing, 2015), pp. 240–251. https://doi.org/10.1007/978-3-319-20267-9_20

85. B. Sun, Q. Ma, S. Zhang, K. Liu, Y. Liu, Iself: towards cold-start emotion labeling using transfer learning with smartphones. ACM Trans. Sensor Netw. **13**(4), 30:1–30:22 (2017). https://doi.org/10.1145/3121049

86. E.R. Thompson, Development and validation of an international English big-five mini-markers. Pers. Individ. Differ. **45**(6), 542–548 (2008). https://doi.org/10.1016/j.paid.2008.06.013

87. R. Wang, G. Harari, P. Hao, X. Zhou, A.T. Campbell, SmartGPA: how smartphones can assess and predict academic performance of college students, in *Proceedings of 2015 ACM International Joint Conference on Pervasive and Ubiquitous Computing (UbiComp 2015)* (ACM, 2015), pp. 295–306. https://doi.org/10.1145/2750858.2804251

88. R. Wang, F. Chen, Z. Chen, T. Li, G. Harari, S. Tignor, X. Zhou, D. Ben-Zeev, A.T. Campbell, StudentLife: assessing mental health, academic performance and behavioral trends of college students using smartphones, in *Proceedings of the 2014 ACM International Joint Conference on Pervasive and Ubiquitous Computing*. UbiComp'14 (ACM, 2014), pp. 3–14. https://doi.org/10.1145/2632048.2632054

89. R. Wang, F. Chen, Z. Chen, T. Li, G. Harari, S. Tignor, X. Zhou, D. Ben-Zeev, A.T. Campbell, StudentLife: using smartphones to assess mental health and academic performance of college students, in *Mobile Health* (Springer, Cham, 2017), pp. 7–33. https://doi.org/10.1007/9783319513942_2

90. W. Wang, G.M. Harari, R. Wang, S.R. Müller, S. Mirjafari, K. Masaba, A.T. Campbell, Sensing behavioral change over time: using within-person variability features from mobile sensing to predict personality traits. Proc. ACM Interact. Mob. Wearable Ubiquit. Technol. **2**(3), 141:1–141:21 (2018). https://doi.org/10.1145/3264951

91. P. Welke, I. Andone, K. Blaszkiewicz, A. Markowetz, Differentiating smartphone users by app usage, in *Proceedings of the 2016 ACM International Joint Conference on Pervasive and Ubiquitous Computing*. UbiComp'16 (ACM, 2016), pp. 519–523. https://doi.org/10.1145/2971648.2971707

92. H. Xiong, Y. Huang, L.E. Barnes, M.S. Gerber, Sensus: a cross-platform, general-purpose system for mobile crowdsensing in human-subject studies, in *Proceedings of 2016 ACM International Joint Conference on Pervasive and Ubiquitous Computing (UbiComp 2016)*. UbiComp'16 (ACM, 2016), pp. 415–426. https://doi.org/10.1145/2971648.2971711

93. R. Xu, R.M. Frey, E. Fleisch, A. Ilic, Understanding the impact of personality traits on mobile app adoption—insights from a large-scale field study. Comput. Hum. Behav. **62**(Supplement C), 244–256 (2016). https://doi.org/10.1016/j.chb.2016.04.011

94. O. Yurur, C. Liu, Z. Sheng, V. Leung, W. Moreno, K. Leung, Context-awareness for mobile sensing: a survey and future directions. IEEE Commun. Surv. Tutor. **18**(1), 1–28 (2014). https://doi.org/10.1109/COMST.2014.2381246

95. X. Zhang, F. Zhuang, W. Li, H. Ying, H. Xiong, S. Lu, Inferring mood instability via smartphone sensing: a multi-view learning approach, in *Proceedings of the 27th ACM International Conference on Multimedia*. MM'19 (Association for Computing Machinery, 2019), pp. 1401–1409. https://doi.org/10.1145/3343031.3350957

96. X. Zhang, W. Li, X. Chen, S. Lu, MoodExplorer: towards compound emotion detection via smartphone sensing. Proc. ACM Interact. Mob. Wearable Ubiquit. Technol. **1**(4), 176:1–176:30 (2018). https://doi.org/10.1145/3161414

97. T. Zhou, Y. Lu, The effects of personality traits on user acceptance of mobile commerce. Int. J. Hum.–Comput. Interact. **27**(6), 545–561 (2011). https://doi.org/10.1080/10447318.2011.555298

Chapter 4
TYDR: Track Your Daily Routine

Our review of the related work for mobile crowdsensing in psychoinformatics shows that there is are no standardized tools or apps for psychoinformatical research. With the Android app TYDR—Track Your Daily Routine—we designed, implemented, and rolled out an app for such psychoinformatical research and analyzed the data we collected (A-Task2). TYDR tracks objective smartphone data measurements, for example, when which app was used. Additionally, TYDR poses psychometric questionnaires, for example, assessing the personality traits of the user, for labeling the unobtrusively collected data. Android allowed for the collection of more types of context data than iOS. We expect that studies conducted with TYDR and their research results are mostly valid in general terms, not only for Android users: research finds that the personality between Android and iOS users differs only a little [22].

Given the lack of a structured approach to privacy awareness, as shown in the related work section, we develop the privacy model PM-MoDaC (Privacy Model for Mobile Data Collection Applications) with nine concrete measures that researchers and software developers can take when implementing apps relating to mobile data collection (A-Task2). Additionally, in the related studies, we observe low sample sizes, often smaller than 100. Based on the data we collected with TYDR, we analyze what patterns we can find regarding the data users are willing to share with researchers (A-Task3).

The remainder of this chapter is structured as follows: TYDR's design and implementation is detailed in Sect. 4.1. We present a general model of context data for smartphone applications in Sect. 4.2. In Sect. 4.3, we detail how the tracking of this sensor data is implemented in TYDR. Section 4.4 gives details about what type of psychoinformatical study TYDR supports. We introduce our privacy model PM-MoDaC in Sect. 4.5. We show the implementation of PM-MoDaC in TYDR in Sect. 4.6. In Sect. 4.7, we perform an extensive statistical evaluation showing what data users are willing to share with researchers, based on the data collected with

F. Beierle, *Integrating Psychoinformatics with Ubiquitous Social Networking*,
T-Labs Series in Telecommunication Services,
https://doi.org/10.1007/978-3-030-68840-0_4

TYDR, while employing PM-MoDaC. Parts of this chapter have been published in
[6, 5, 7, 4].

4.1   TYDR Design and Implementation

There are several things to consider when developing an app for mobile crowdsensing. The more users we can attract, the more reliable results about the relationship between smartphone data and the user's personality will be. The app should have an appealing interface and should offer some attractive feature for the user. As we are developing for a mobile device, we have to consider the restrictions these devices pose, like battery and space limitations. Additionally, dealing with sensitive user data, privacy awareness should be considered. We cover privacy aspects separately in Sect. 4.5.

We implemented TYDR natively for Android in Java, using SQLite as a local database. Our custom backend utilizes Loopback,[1] a NodeJS-based framework, which offers REST interfaces, at which the app registers and then uploads data. In the following, we detail how TYDR makes the data tracking its core feature by processing and visualizing it for the user. The three main components of TYDR's user interface are the main screen, the questionnaires the users can fill out and their results, and the permanent notification.

4.1.1   Main Screen

Figure 4.1a shows the main screen after starting TYDR for the first time. The tile-based design gives the user an immediate overview of the data for the current day. The gray overlay with the "Grant Permission" buttons indicates missing permissions that the user has to give in order to see more statistics. Each tile can be touched to slide open a bigger tile with a more detailed view of the data. Figure 4.1b shows more detailed information on location data after touching the corresponding tile. The visited locations are visualized on a map. The staypoints, i.e., staying more than 20 min at one place, are indicated by a marker. Pressing the "Full Map" button shows the full screen view of the day's location history and related data like weather (see Fig. 4.1c). Figure 4.1d shows the usage times for the past week. This tile is shown after the small "Usage" tile on the main screen is touched. Similarly, Fig. 4.1e shows the number of photos taken, distinguished by front and back camera. Figure 4.1f shows that using a calendar, old data from previous days can be accessed. This functionality is accessed from the top right icon of the main screen. Not shown in

---

[1] https://loopback.io/.

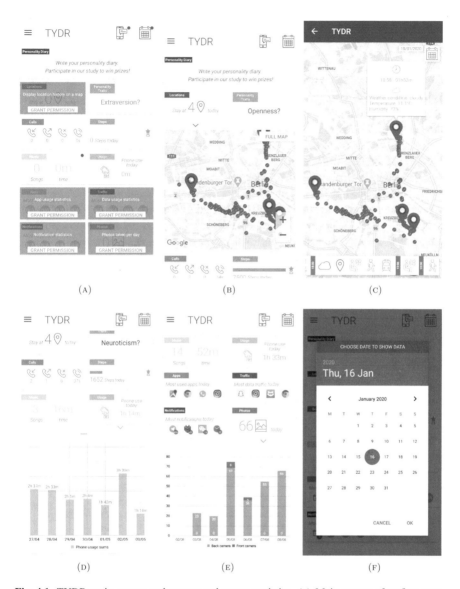

Fig. 4.1 TYDR main screen and sensor and usage statistics. (**a**) Main screen after first start. (**b**) Location data. (**c**) Location and weather data in full-screen view. (**d**) Usage times. (**e**) Photo statistics. (**f**) Calendar to choose what data to see

these screenshots are additional tiles related to call statistics, music statistics, steps taken, the most used apps, apps with most traffic, and the number of notifications per app. The related statistics are visualized in a similar way as the shown examples.

4.1.2 Mobile Questionnaire

TYDR utilizes questionnaires for demographic data that are not possible to track automatically. Additionally, we use standardized psychometric questionnaires to assess the user's personality with which we label the collected smartphone data. To be able to update the questionnaires independently from updating the whole app, the latest questionnaire version is fetched from the backend.

To the best of our knowledge, currently there is no official or widely adopted library for mobile questionnaires on the Android platform. Following general mobile survey design guidelines,[2] we developed a questionnaire UI (user interface). Only one question is displayed at a time, which avoids scrolling. The users can switch between apps or turn off the screen and continue where they left off when resuming TYDR. A progress bar indicates how much of the current questionnaire is already filled out. The incentive for the user to fill out the personality questionnaires is to see their results in the related tile.

Figure 4.2a shows the questionnaire interface. Figure 4.2b shows the extended tile shown after completing the questionnaire and touching the "Personality Traits"

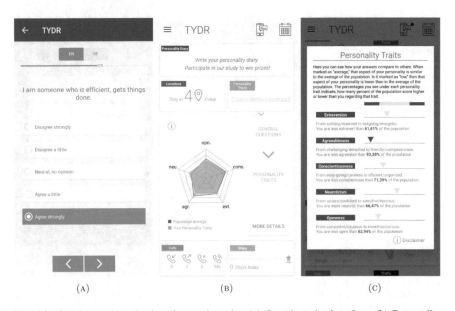

(A)	(B)	(C)

Fig. 4.2 TYDR questionnaire interface and results. (**a**) Questionnaire interface. (**b**) Personality traits questionnaire results—extended tile. (**c**) Personality traits questionnaire results—detailed view

---

[2]E.g., https://www.surveymonkey.com/mp/how-to-create-surveys/ or https://www.uxmatters.com/mt/archives/2017/02/8-best-practices-for-mobile-form-design.php.

tile. Tapping on the "More Details" button opens a full-screen view with more details about the user's personality (see Fig. 4.2c).

4.1.3   Customizable Permanent Notification

In order to improve the chances of TYDR not being stopped by task cleaner apps, we implemented a permanent notification. It runs those services in the foreground that track data with a high frequency. To make it appealing, we followed the same approach as for the main screen: show the user meaningful and informative figures based on processed tracking data.

The permanent notification will show up in the notification bar and the lockscreen. The notification is designed to be adaptive to the user's interests by offering the possibility to configure what information is displayed (see Fig. 4.3). The *Preview* section in the figure shows what the notification will look like.

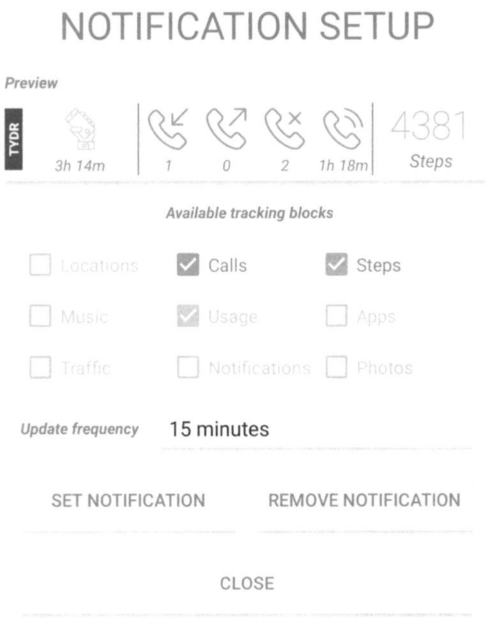

Fig. 4.3   The user can configure the notification. The *Preview* section shows what the notification will look like

4.2 Context Data Categories

In order to get an overview of what context data a crowdsensing app for psychoinformatical studies can collect as objective measurements, we develop a general context data model for smartphone applications. We present our model in Table 4.1. The four context data categories are *physical* conditions and activity, *device* status and usage, *core functions* usage, and *app* usage. Furthermore, an additional technical category constitutes the explicit permission by the user in order to allow an app to access data from the given source. This has important implications, e.g., for answering the question if it is possible to develop a library for personality prediction that does not require explicit permissions.

Physical conditions and activity deal with the physical context of the user that is not related to the interaction with the smartphone. Here, sensors deliver data without the user interacting with the phone, e.g., location or taken steps. The ambient light sensor typically offers data only when the screen is active, so when the user is interacting with the phone. However, as the light sensor's data is related to the physical context, i.e., the light level of the environment of the user, we regard it as part of the *physical* category.

Table 4.1 Context data model for the categorization of context data for smartphone applications. The last column indicates if an explicit user permission is required (Android)

	Category				Permission
	Physical	Device	Core functions	Apps	
Location	•				•
Weather	•				(•)
Ambient light sensor	•				
Ambient noise level	•				•
Accelerometer	•				
Gyroscope	•				
Activity	•				
Steps	•				
Screen and lock state		•			
Headphone unplug/plug		•			
Battery and charging		•			
WiFi		•			
Bluetooth		•			
Calls metadata			•		•
Music metadata			•		(•)
Photos metadata			•		•
Notifications metadata			•	•	•
App usage				•	•
App traffic				•	•

The category *device* status and usage designates data that is related to the status and the connectivity of the smartphone. This comprises screen/lock state, headphone connection status, battery level, and charging status as well as WiFi and Bluetooth connectivity.

Core functions usage deals with the users' interaction with core functionalities of the phone, regardless of which specific apps they are using for it. The core functions comprise calling, music listening, taking photos, and dealing with notifications.

The fourth category is *app* usage, dealing with data about the usage and traffic of specific apps. Notifications fit both in the *core functions* and the *apps* categories because they can be related to either.

The *permission* column is based on the permission system introduced with Android 6.0 (API 23). Weather is given in parenthesis because it can only be collected if the location is available, so it is bound to the location permission. Music is given in parenthesis as well. Most major music player apps or music streaming apps automatically broadcast metadata about music that the user is currently listening to. The broadcast events can be received by any app that subscribes as a listener. However, for Spotify, such broadcasting has to be activated manually.

There is additional data that could be gathered from mobile phones, e.g., touch patterns or touch intensity (cf., e.g., [11]). However, data points are only available when the developed app itself is in the foreground, not whenever any other app is being used. Thus, it might be difficult or impossible to get meaningful data for purposes like personality prediction.

In general, our context data model can be helpful for the development of any context-aware service, e.g., in the areas of ubiquitous computing and social networking (cf. Part II). After collecting data, we can analyze to what extent the quality of the context data varies between the different available Android devices. Our context data categorization allows to address different specific questions. Take the prediction of personality from context data, for example. Then, specific questions are, for example, whether the physical context alone can predict personality, how meaningful metadata is, or how accurate the prediction can be if the user did not give any explicit permissions.

4.3 Context Data Tracking

Google released the Google Awareness API[3] that offers developers to retrieve different context data through one API (time, location, places, beacons, headphones, activity, weather).[4] There are two ways of retrieving data: *Fence API* and *Snapshot*

---

[3] https://developers.google.com/awareness/.

[4] After the release of TYDR end of 2018, Google made the *place* a paid feature available via the Places API end of 2019 and removed the *weather* context data point without offering an alternative end of January 2020.

API. A snapshot yields current data from the seven sources. Through the Fence API, the developer can register listeners and receives a callback when the desired conditions are met. These two approaches are useful for the development of context-aware applications that either instantly need the current context of the user or want to be notified when the user is in a specific context. Aiming at tracking the user's context, we have to go beyond what the Google Awareness API offers. In TYDR, we track location, weather, ambient light sensor, accelerometer, activity, steps, phone unlock/lock, headphone unplug/plug, battery and charging, WiFi, Bluetooth, calls metadata, music metadata, photos metadata, notifications metadata, app usage, and app traffic. This encompasses all of the data captured in our context data model (cf. Sect. 4.2) except for gyroscope and microphone. For accelerometer and gyroscope, capturing very high-frequency data would quickly exceed space and maybe battery limitations. We decided to track a limited amount of accelerometer data. Regarding data from the microphone—for example, to assess the ambient noise level—we decided not to track it. We anticipated that users would be very reluctant to let a mobile data collection app access their microphone. Even when immediately processing the data from the microphone and only storing a single noise level value, the user still has to trust that is what the app actually does.

For some of the listed data sources we want to track, a passive, listener-based approach is possible. For example, the music that is played is broadcast by (most) music player apps. We can just register a listener and track the played back music whenever the user is listening to it. The same approach can be used for tracking the usage of the phone: registering listeners for locking and unlocking events enables us to track when the user was (most likely) interacting with the phone.

Besides such a listener-based approach, in some cases, we have to do periodical tracking. Given the related permission, the Android system offers information about which apps were used for what number of seconds. Android also offers information about the traffic each app caused. Such data about the user are especially meaningful when we have it on a fine-granular level. For example, having one data point per week that indicates for how long a user used a specific app would probably yield less interesting insights than knowing the app usage duration for each hour. As there are no listeners available for querying such app statistics, it is necessary to schedule periodic queries.

For periodically tracking context data, there is the trade-off between the:

1. amount of data we have to store,
2. battery consumption, and
3. level of detail of the information.

In the most extreme case, the frequency with which we query context data is very high, in order to increase the level of detail for quickly changing data. Then we—and the user—have to accept a higher battery consumption. An example for this quickly changing data is that of the light sensor or the accelerometer. We did some further optimizations, for example, by only tracking accelerometer data when we detect movement indicated by the step count we track. Although the light sensor only yields data when the phone is unlocked, the frequency with which it yields data is still so high that it would use too much space. Furthermore, such detailed light sensor data would not be that useful for our research purpose. What could be interesting is to check whether the user is in a dark or in a well-lit place. To reduce the amount of data we store from the light sensor, we divided the possible range of light sensor values into several segments and only store changes between the segments. To counteract the effects of rapid changes between the limits of the defined segments, we implemented a hysteresis. The introduced inaccuracy is negligible for our research purpose.

4.4   Study Design

With TYDR, we want to enable studies that are related to psychological concepts that are rather static as well as aspects that are more dynamic and can change over time (cf. Sect. 3.3). As a rather static aspect, we query the user with a personality traits questionnaire. For this, we use the Big Five Inventory 2 (BFI-2) questionnaire [15, 41]. It consists of 60 items. Each of the 5 traits is assessed with 12 items based on a 5-point Likert scale, ranging from 1 (strongly disagree) to 5 (strongly agree). Moreover, the expression of personality traits fluctuates within persons across time [21]. For example, a person who scores high on neuroticism will experience negative mood more often than other people but may still vary considerably in the experience of negative mood across time, e.g., depending on situational circumstances. This within-person variability of emotions and behaviors is captured by the term *personality states*. In order to register those aspects, we utilize the PDD (Personality Dynamics Diary) questionnaire, which captures the user's experience of daily situations and behaviors [49]. The PDD questionnaire could be replaced with other questionnaires assessing, e.g., mood or stress, in order to enable other EMA (ecological momentary assessment, cf. Sect. 3.3) studies.

4.5   PM-MoDaC: Privacy Model for Mobile Data Collection Applications

Surveying related work in Sect. 3.3, we observe a lack of privacy awareness in psychoinformatical apps related to the mobile collection of data. Overall, we observe that if there is information given about privacy protection in related work, it is typically not very detailed and usually only covers some of the aspects given in the following privacy model. With PM-MoDaC, we introduce our privacy model comprising nine concrete measures that can be implemented. By adding additional measures, this model could be extended to fulfill additional requirements for regional laws or special use cases like certifying an app as a medical product. In this section, we present a comprehensive overview of concrete measures that can

Table 4.2 The nine privacy
measures of PM-MoDaC

(A)	User consent
(B)	Let users view their own data
(C)	Opt-out option
(D)	Approval by ethics commission/review board
(E)	Random identifiers
(F)	Data anonymization
(G)	Utilize permission system
(H)	Secured transfer
(I)	Identifying individual users without linking to their collected data

be taken to protect user privacy. To the best of our knowledge, we are the first to provide such a comprehensive privacy model for applications related to mobile data collection.

Our **Privacy Model for Mobile Data Collection Applications (PM-MoDaC)** comprises nine privacy measures (PM) given in Table 4.2. Inside the EU, the GDPR (General Data Protection Regulation) regulates the processing of personal data and thus applies to many researchers and participants of studies. While there are uncertainties regarding the specifics on how to comply to those regulations [20], some of the requirements posed by the GDPR are rather concrete. "Informed consent" is the basis that is necessary for any data processing [26] and is covered by PM A. Users have to have access to their data in a "portable format" [26], which is covered by PM B. Opting out has to be possible at any point in time [26], which is covered by PM C. The GDPR specifically allows archival storage for scientific research [26]. When designing applications, "state-of-the-art technical and organizational measures" have to be taken, although what exactly that means is not explicitly defined [26]. In the following, we describe the concrete measures we propose.

(A) User Consent   Before installing the app, the user should be explained what data exactly is being collected and for what purpose. These are typical aspects covered in a privacy policy that the user has to agree to before using an app. The aspect of *user consent* is mentioned in [27, 39, 17, 40, 32, 42, 33].

(B) Let Users View Their Own Data   [27] discuss this aspect of privacy protection. By letting the users see the data that is being collected, they can make a more informed decision about sharing it. Furthermore, a user should be able to export all the data stored about him/her.

(C) Opt-out Option   The possibility of opting out is rarely mentioned (e.g., [45, 39]). Especially after viewing their own data (see previous point), users might decide that they no longer want to use the app or participate in the study.

(D) Approval by Ethics Commission/Review Board   Psychological or medical studies typically require prior approval by an ethics commission or review board. Some of the related works explicitly state that such approval was given for their studies [9, 19, 25, 2, 43, 36, 1, 35, 32, 37, 24, 23]. This aspect of privacy protection is more on a meta-level, as an ethics commission/review board might check the other points mentioned in this privacy model.

(E) Random Identifiers When starting an app, often a login is required. This poses the privacy risk of linking highly sensitive data with personal details, e.g., the user's Facebook account details if a Facebook account was used to log in. A few related studies describe using random identifiers [45, 47, 34, 34, 14, 18, 44]. This point relies on the type of study being conducted. Investigating the relationship between collected sensor data and, for example, the number of Facebook friends would probably require the user to log in via Facebook. On a technical level for the Android system, an ID provided by the Google Play Services proved itself suitable as a random ID.

(F) Data Anonymization This aspect is mentioned most commonly in the related work [27, 12, 13, 30, 19, 25, 34, 43, 1, 35, 39, 2, 44, 10, 46, 48, 8, 38]. If details are given, they usually describe how one-way hash functions are used to obfuscate personally identifiable data like telephone numbers, WiFi SSIDs, or Bluetooth addresses.

TYDR only stores clear text data where it is necessary for the research purpose. Our context data model from Sect. 4.2 helps to analyze why metadata will suffice in most cases. Consider notifications, for example. Depending on the application, they might contain highly sensitive data, e.g., the message content of a messenger application. The content of the notification is not relevant for our research purpose. The app name that caused the notification however is, as one could easily imagine a relationship between, e.g., the personality trait *extraversion* and the frequency of chat/messaging app notifications.

An additional point to consider regarding data anonymization is where the anonymization happens. In [25], the authors describe how the anonymization is taking place on the backend that the data is being sent to, before being stored. In TYDR, the anonymization process is taking place on the device itself, before storing to the local device and before sending data to the backend. The backend consists of server, application, and database. So, even if our backend was compromised, the attacker would only be able to access data that is already anonymized.

(G) Utilize Permission System This point is specifically related to the Android permission system that was introduced with Android 6.0 (cf. Sect. 4.2). By itself, it can already make the users more aware of what data/sensor is being accessed by an application. The designers of an application still have influence over how they make use of the system though. Requesting all permissions at the first start of an app, e.g., gives the user little insight about what each permission is used for. Instead, the app should request a permission at the point where it is needed and explain to the user what the accessed data source is being used for.

(H) Secured Transfer The point of having secured data transfer between mobile device and backend is explicitly mentioned in a few of the related works [45, 19, 1, 34, 18, 46, 8, 38]. An alternative way is to only locally collect data and ask users in a lab session to bring their phone and copy the data then. Such an approach would severely limit the possible scope of a study.

(I) Identifying Individual Users Without Linking to Their Collected Data In psychological studies, it is common that users are compensated with university course credits, get paid to participate, or have the chance to win money/vouchers in a raffle after study completion. In order to contact the study participants, contact information is needed, which might contradict PM E. In order to alleviate this concern, we developed a process that allows us to both:

1. check if users successfully participated in a study and
2. contact study participants without knowing what smartphone data belongs to them.

This way, we can directly draw and contact the winners of a raffle, without having a link between collected data and meaningful user identifiers, i.e., email addresses. Additionally, generated participation codes can further be used for claiming university credit points, if applicable. There are three steps to our design of this system:

User sign-up. In order to participate in the study, additionally to just installing the app, users have to sign up with an email address, so we can contact the winners of the raffle. Entering an email address has the potential to de-anonymize the data. We store the data about the study participants in a separate table in the backend that is not linked to the tables containing smartphone data.

Checking study participation success. The planned study includes the commitment of the participants to fill out daily psychological questionnaires. Users who do not fill those out regularly should not be eligible to receive compensation. As we have no link between the identifiable study participants and their smartphone data, we cannot check the rate of filling out the daily questionnaire on the backend. However, we can check this rate in the app and report to the backend whether it meets the required percentage.

Generating participation codes. After successful study completion, the user can trigger the study completion by pressing the related button in the sidebar menu. This triggers the app to let the backend know about the successful completion of the study. The backend then generates an individual participation code and replies to the app's query with it. This code can then be used to indicate successful participation to, e.g., claim university credit points.

To summarize, the process consists of storing contact data separately from the collected smartphone data and letting the app check the requirements for successful study completion, in our case the daily completion of the PDD questionnaire. This way, we can create incentives for users to install and use the app while simultaneously preserving user privacy.

In [29], the authors also deal with the often contradicting requirements of privacy and user incentivization. Here, the authors propose a system with automatic payouts. In this case, technological advancement is faster than bureaucratic processes, which sometimes still require manual approval or handwritten signatures.

4.6   Implementation of PM-MoDaC in TYDR

With TYDR, to the best of our knowledge, we are the first to implement a privacy model comprising all nine privacy measures listed in Sect. 4.5. The visualization of the data that is collected about the user is TYDR's core feature (PM B). The ethics commission of Technische Universität Berlin approved the use of TYDR in a psychological study (PM D). Users can opt out via the contact form from the sidebar menu (PM C). In Fig. 4.4, we show a diagram of the main processes in the TYDR app. The person icon in a process signifies that the user is actively doing something. All other processes are part of the app and do not require user interaction. Starting TYDR for the first time, the user has to confirm the terms and the privacy policy (cf. PM A). Only then the five processes of the app are started. Note that there is no login process; the system uses a random unique identifier (PM E).

At the bottom of Fig. 4.4, we show that the app starts the data collection (Process 4). The data collection engine already anonymizes the data before storing it (PM F). The uploading via a secure connection (PM H) is started after the app registered itself with the backend. The upload process is repeated every 24 h (Process 5).

The process at the top of the figure shows the main menu of TYDR (Process 1) (also cf. Fig. 4.1a). From here, the user can grant permissions (cf. Table 4.1 and Fig. 4.1a; PM G), which influences the data collection. The user can also fill out the general (demographic information) and the personality traits questionnaire (*Personality Traits* tile in Fig. 4.2). TYDR offers a permanent notification, displaying information on the lockscreen and the notification bar (Process 2). The data to be displayed can be configured by the user via the second icon from the right at the top (Fig. 4.1a). The tracking of personality states via the PDD questionnaire is designed to be optional (Process 3). Configuring the PDD questionnaire via the *Personality States* tile (Fig. 4.1a), the user can (de-)activate this feature. In order to collect data

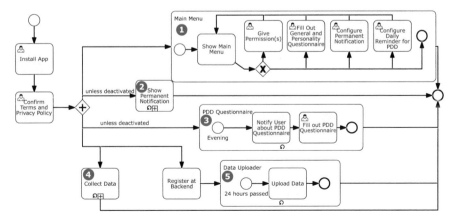

Fig. 4.4 Process diagram for our smartphone sensor and usage data tracking app TYDR

labeled with personality states, we conducted a study where users commit to turning this feature on for a certain period of time. The registration for this study takes into account PM I.

4.7   Evaluation of PM-MoDaC

In order to gain insights about mobile data collection apps and the proposed privacy model, we released TYDR on Google Play in October 2018, started advertising mid-November 2018, and collected the data this evaluation is based on. In order to advertise TYDR, we printed flyers, advertised TYDR in lectures,[5] and posted on social media. Most TYDR installations happened after a German website that deals with Android-related content reported about new apps including TYDR. Looking at PM-MoDaC, there are some aspects that the user cannot interact with, i.e., PM B, D, E, F, H, and I. The user can interact with the remaining three aspects, PM A (user consent), C (opt-out option), and G (permission system). To be more precise, the user can accept or decline the terms and conditions and the privacy policy (PM A). The user can choose to opt out and request his/her data to be deleted (PM C). Furthermore, the user can determine which permissions he/she grants to TYDR (PM G). This section addresses research task A-Task3: *Research what data users are willing to share with researchers and how the sharing of data relates to user characteristics and personality traits.*

Datasets   For the 2-month period between 12 November 2018 and 17 January 2019, Google reports 3010 installations. In our database, we have 2876 users, and, after data cleaning, we have 1560 users with valid information about their permissions settings. Dataset DS1 contains these users. Out of these users, 634 filled out a demographic questionnaire (dataset DS2). Out of those, 461 also filled out the Big Five personality traits questionnaire (dataset DS3). Note that this makes DS3 a subset of DS2 and DS2 a subset of DS1: $DS3 \subset DS2 \subset DS1$. In Table 4.3 and Fig. 4.5, we give an overview about the datasets and how the users interacted with the permission system (PM G).

For DS2, demographic information is available. We observe that 83% (527) of TYDR users are male, while 17% (107) are female. The mean age of the TYDR users is 33.93 years (SD: 12.43). For the users in DS3, we provide the average for the Big Five personality traits scores in Table 4.4. We compare the values with the population averages given in [15] and give a standardized effect size (Cohen's d). The biggest difference in comparison with the population average is that TYDR users score higher on the trait *openness to experience*. We assume that TYDR by its nature as an app that displays smartphone statistics might attract users with such a personality pattern. Considering the ubiquity of smartphones, we assume that the

---

[5]At Technische Universität Berlin, Berlin Psychological University, University of Kassel, University of Ulm, University of Regensburg, University of Zurich, and Danube University Krems.

Table 4.3 Datasets used in the evaluation. About the users in DS1, we have valid permission information. The users in DS2 filled out the demographic questionnaire, while users in DS3 also filled out the personality questionnaire; hence we have DS3 $\subset$ DS2 $\subset$ DS1

Granted permissions	DS1	DS2	DS3
All permissions	660 (42%)	387 (61%)	287 (62%)
Only some permissions	344 (22%)	162 (26%)	123 (27%)
No permissions	556 (36%)	85 (13%)	51 (11%)
Sum	1560	634	461

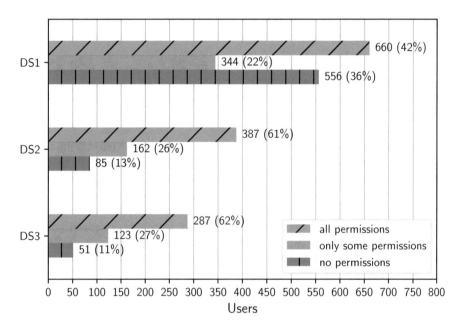

Fig. 4.5 Visualization of the datasets DS1, DS2, and DS3; see Table 4.3

Table 4.4 Average Big Five personality trait scores (scale 1–5) of TYDR users (DS3) in comparison with the population averages given in [15], including the differences between the means and Cohen's d effect size

Pers. Traits	TYDR			Population average			Diff.	Cohen's d
	Mean	SD	Cronbach's Alpha	Mean	SD	Cronbach's Alpha		
Openness to exp.	3.70	0.63	0.78	3.38	0.64	0.84	+0.32	+0.50
Conscientiousness	3.37	0.64	0.81	3.67	0.62	0.87	−0.30	−0.48
Extraversion	3.16	0.67	0.82	3.22	0.63	0.86	−0.06	−0.10
Agreeableness	3.64	0.54	0.72	3.76	0.51	0.81	−0.12	−0.24
Neuroticism	2.85	0.82	0.89	2.72	0.67	0.88	+0.13	+0.19

average smartphone user resembles the average of the population. The results shown in Table 4.4 could then be evidence that the user base acquired by an app depends on the type of the app.

(A) User Consent There is a difference of 134 users between the number of installations reported by Google (3010) and the number of users in our database (2876). Only after consenting to the terms and conditions as well as the privacy policy, a connection to our backend is established (cf. Fig. 4.4). After the connection is established, an entry in our database is created. This yields a rate of 95.5% of users who accepted our terms and conditions as well as our privacy policy. The remaining 134 users (4.5%) could have never opened the app after installation, there could have been an error during the connection with our backend, or they had concerns about the given terms and policy.

We have data about 778 users (27% of the 2876 users registered with our backend) who confirmed terms and policy and gave the permission to access their app usage statistics. For those users, we can see how long they spend on the *Welcome* activity of TYDR. This activity is shown to the user when he/she starts the app for the first time. It consists of two screens (see Fig. 4.6). The first shows the TYDR logo and contains a brief description of what TYDR does (cf. Fig. 4.6a). On the second screen, the terms and conditions and the privacy policy are displayed; they have a combined length of overall around 1300 words (cf. Fig. 4.6b). The median time spent in this activity is only 10 s (SD: 23.6), clearly not enough to read the texts of the privacy policy and the terms and conditions. Figure 4.7 shows a histogram of the time users spent in the *Welcome* activity. We observe that, 90% (700/778) spent less than 30 s in the activity. Only 3.5% (27/778) spent more than 1 minute, while 1.1% (9/778) spent more than 2 min, with the longest time spent by any user being 258 s (4.3 min). Assuming an average reading speed of 200 words per minute, reading the full texts would take about 6.5 min. While it is possible to (re-)read both texts at any later point from within a menu in the app, our data confirms the stereotype of the user not reading the privacy policy and blindly confirming. We assume that those with high concerns about sharing data will not have installed TYDR in the first place. The design and display of privacy policies and terms and conditions remains a challenge. There are contradicting requirements of providing a simple and quickly graspable overview and providing all technical details. Overall, likely, convenience trumps over privacy concerns when it comes to reading the texts.

(C) Opt-Out Option Only one user withdrew his or her data from the study. According to the GDPR, each service has to offer the option to delete all of a user's data. In that respect, this is not a feature unique to TYDR but should be a feature available in any service (offered in the EU). In TYDR, we explicitly described the opt-out option in the privacy policy. However, we did not implement a dedicated button for this. This could have created a bias toward users not requesting the deletion of their data.

(G) Permission System In the following, we will look into the users' characteristics captured by the questionnaires and set those in relation to which permissions

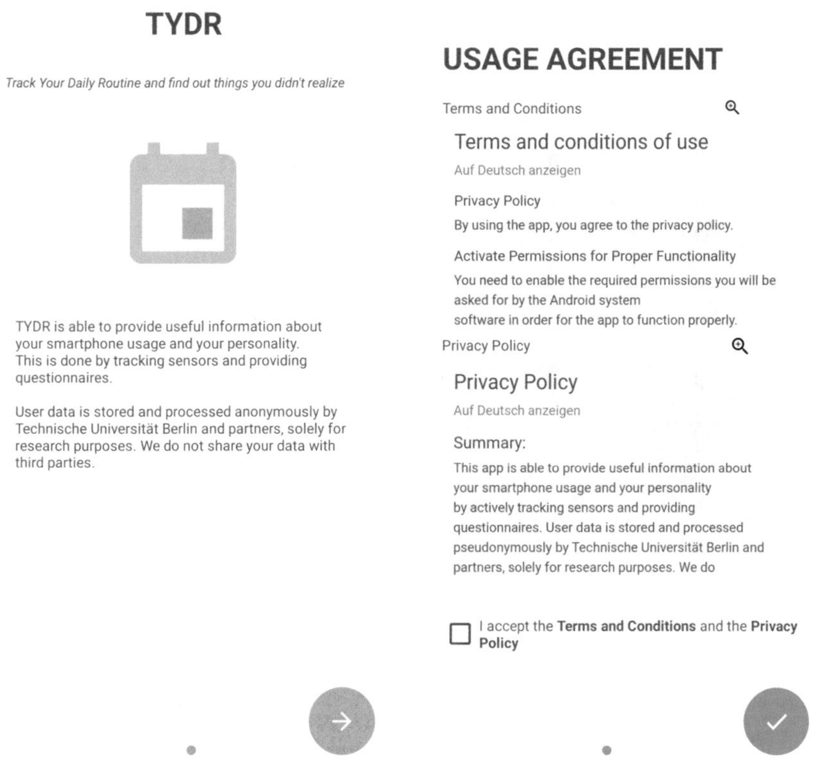

TYDR

Track Your Daily Routine and find out things you didn't realize

TYDR is able to provide useful information about your smartphone usage and your personality. This is done by tracking sensors and providing questionnaires.

User data is stored and processed anonymously by Technische Universität Berlin and partners, solely for research purposes. We do not share your data with third parties.

USAGE AGREEMENT

Terms and Conditions

Terms and conditions of use

Auf Deutsch anzeigen

Privacy Policy

By using the app, you agree to the privacy policy.

Activate Permissions for Proper Functionality

You need to enable the required permissions you will be asked for by the Android system software in order for the app to function properly.

Privacy Policy

Privacy Policy

Auf Deutsch anzeigen

Summary:

This app is able to provide useful information about your smartphone usage and your personality by actively tracking sensors and providing questionnaires. User data is stored and processed pseudonymously by Technische Universität Berlin and partners, solely for research purposes. We do

☐ I accept the **Terms and Conditions** and the **Privacy Policy**

(A)   (B)

Fig. 4.6 TYDR Welcome activity. (**a**) TYDR Welcome screen. (**b**) TYDR Usage Agreement screen

they gave. This can give us insights into what concern the users have regarding which data source.

The permissions used in TYDR are:

- storage
- location
- call log
- app usage
- notifications

For the last two, Android opens an extra confirmation page for the user because apps that are granted these permissions can access especially sensitive data.

To reduce complexity, in the following, we looked into users that granted either all or no permissions. To test for demographic and psychological differences between TYDR user groups giving all or no permissions, we compared permission

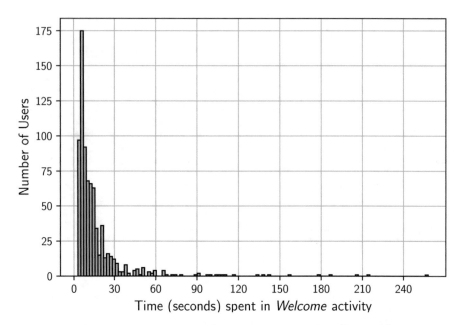

Fig. 4.7 Histogram of time (seconds) spent in TYDR's *Welcome* activity, based on 778 users

groups with regard to age, gender, education, and personality traits and facets. For
the comparison of the mean age and personality traits and facets scores between
groups, we applied Student's t-tests, whereas Wilcoxon rank-sum tests were used
to test for differences in education. Finally, a chi-square test was used to test for
differential distributions between the genders. The statistical tests performed were
two-tailed, and the significance level was set to $p < 0.05$. All analyses were
conducted using R.

There are some aspects to consider when interpreting the results: users who
did not fill out the demographic or personality questionnaire because of privacy
concerns are not reflected in the following statistics, simply because their data is not
available. Filling out the questionnaires itself could be related to concerns about
sharing data as well, i.e., a user could decide not to answer the questionnaires
because of privacy concerns. The difference to the automatically collected context
data is that the demographic and personality data is static and not updated.
Furthermore, not giving permissions does not necessarily mean the user had privacy
concerns; there could be other reasons which we will mention in the following.

Note on DS1 Overall, 30% (471) of our users did not fill out any questionnaire
and did not give any permission (i.e., the cardinality of {*user* in DS1 \ DS2 |
user did not give any permissions}, cf. Table 4.3, is 471). We assume this group
of users might not have had any interest in using the app after starting it.

Table 4.5 Gender and permission settings (DS2)

Granted permissions	Female	Male	Sum
All permissions	51 (48%)	336 (64%)	387 (61%)
Only some permissions	32 (30%)	130 (25%)	162 (26%)
No permissions	24 (22%)	61 (12%)	85 (13%)
Sum	107	527	634

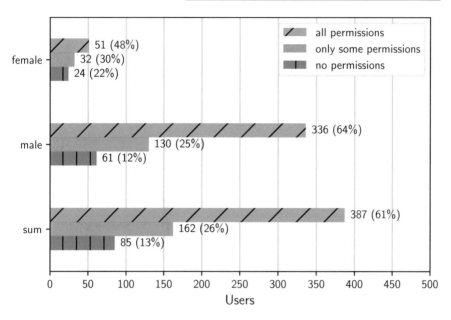

Fig. 4.8 Visualization of gender and permission settings (DS2) (see Table 4.5)

Gender The chi-square test indicates that there is a relationship between gender and having given no or all permissions ($p = 0.001$). We observe that 12% of our male users did not give any permissions, while 22% of the female users did not give any permissions (see Table 4.5 and Fig. 4.8).[6] On the other hand, 64% of male users granted all permissions, while only 48% of female users granted all permissions. This could be evidence that female users are less willing to share their data. An alternative explanation could be that the gender difference might also be based on a difference in interest: maybe some female users installed the app, had a look at it, and decided they do not want to use it because they are not interested.

Age The mean age of the users giving no permissions is lower than that of those giving all permissions (29.75 years vs. 34.87 years; $p = 0.0002$). The violin plot in Fig. 4.9 shows the age distribution for each user group giving all or no permission,

---

[6]The sums add up to 100%; filling out the demographic questionnaire, users had to specify their gender.

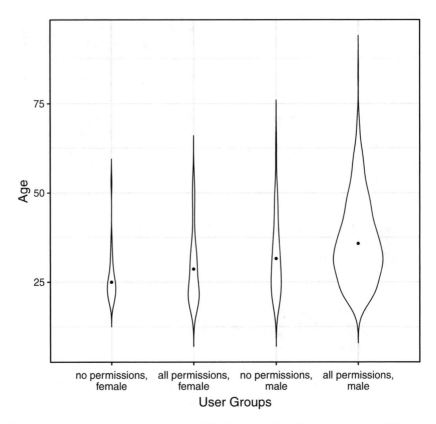

Fig. 4.9 Age distribution showing the age of TYDR users giving all or no permission. The size of the plot indicates the sample size and the dot indicates the mean (DS2)

divided by gender. The size indicates the sample size and the dot gives the mean for the user group. We observe that TYDR's female users are younger on average. For both male and female users, the average age is lower in the group of those users giving no permissions. There are a few ways to interpret these results. The younger the people are, the more likely it is that they grew up with smartphones from an earlier age. This group of users might be more prone to installing and quickly uninstalling apps without using every feature. Another interpretation could be that younger users tend to have more technical background knowledge and tend to have a heightened sensitivity toward data privacy. If our observation is related to privacy concerns, it is consistent with [31], in which the authors found that young people (16–25 in this case) are most concerned about privacy issues.

Education In our demographic questionnaire, we asked for the highest completed level of education. We did not find any statistically significant association between level of education and giving all or no permissions.

Table 4.6 Big Five
personality traits and facets
scores for TYDR users giving
all or no permissions (DS3)

Pers. traits and facets	All Perm.		No Perm.	
	Mean	SD	Mean	SD
Openness to experience	3.67	0.64	3.83	0.64
Aesthetic sensitivity	3.83	0.91	3.55	1.06
Creative imagination	3.71	0.81	3.72	0.83
Intellectual curiosity	3.96	0.75	4.23	0.60
Conscientiousness	3.39	0.63	3.36	0.60
Organization	3.42	0.90	3.36	0.90
Productiveness	3.22	0.83	3.15	0.72
Responsibility	3.53	0.69	3.57	0.73
Extraversion	3.14	0.65	3.33	0.80
Assertiveness	3.23	0.80	3.50	0.89
Energy level	3.34	0.80	3.48	0.98
Sociability	2.83	0.88	3.00	0.97
Agreeableness	3.62	0.53	3.67	0.63
Compassion	3.73	0.70	3.82	0.79
Respectfulness	3.90	0.64	3.86	0.78
Trust	3.23	0.74	3.32	0.72
Neuroticism	2.88	0.81	2.92	0.86
Depression	2.74	0.99	2.74	0.98
Anxiety	3.18	0.89	3.24	0.92
Emotional volatility	2.73	0.96	2.80	1.07

Personality We do not observe any difference with respect to personality traits of
the user and giving all or no permissions (all p values > 0.1). With the personality
questionnaire used, the five personality dimensions can each be separated into three
facets. Looking into those, we observe that the facet *intellectual curiosity*, part of
the trait *openness to experience*, is higher in those who did not give any permissions
($p = 0.005$). However, this does not necessarily mean that intellectually curious
people are less willing to share their data or have more privacy concerns. It could
be that the group of users that filled out the personality questionnaire but did not
grant any system permissions consisted of users who were more curious about their
personality profile and less about their smartphone usage statistics. In Table 4.6, we
present the means and standard deviations of the personality traits and facets scores
for the two TYDR user groups giving all and no permissions.

Discussion/Results In [16], the authors surveyed patients if they would be willing
to use an app related to assessing their mental health disorder. While most were
willing to install, willingness to give certain permissions was lower. The authors
report that 68% of the surveyed patients may be willing to agree to have the state of
their screen monitored by such an app. At least on Android, most likely, users might
not even notice which apps are already doing this, as, in order to do so, no explicit
permission is required from the user (cf. Table 4.1). On average, the surveyed people
were least willing to grant access to share audio recordings and SMS content—

content TYDR does not collect in anticipation of such unwillingness. In contrast to using a survey, we performed analyses on data collected with an actually deployed app. Furthermore, the target audience in our case was very broad; users might have different concerns about apps relating to mental health. In DS2,[7] only 26% of users granted only some permissions, while 61% granted all of them (cf. Table 4.5). This could indicate that users might actually be willing to grant more permissions than they might indicate in a survey.

In general, not granting certain permissions to an app could relate to privacy concerns or concerns about data sharing but could also just be evidence for a lack of interest in the app. Overall, we feel we found quite a lot of users to try and use TYDR. Almost all of the feedback we received from our users was about certain features and not related to privacy or data usage.

From this and from the data analysis we conducted, we found evidence for the following points that address A-Task3 (*Research what data users are willing to share with researchers and how the sharing of data relates to user characteristics and personality traits.*):

- PM-MoDaC seems to be a valid approach to dealing with privacy in mobile data collection apps.
- Younger users tend to be more concerned about privacy/data sharing.
- Female users tend to be less willing to share data.
- Education does not seem to be a factor related to data sharing.
- Personality traits do not seem to be a factor related to data sharing.
- Depending on the type of app, the user base might be biased toward certain personality traits.

We consider these aspects as being helpful for other researchers who conduct studies with mobile applications. More specifically, our revealed aspects can be seen as indicators to what kind of data can be expected in this context. Our findings confirm the role of age and gender and willingness to click on recommendations [3, 28]. Younger users and female users were less likely to click on recommendations. This shows that our findings fall in line with research in other domains, beyond the scope of psychoinformatics.

References

1. I. Andone, K. Błaszkiewicz, M. Eibes, B. Trendafilov, C. Montag, A. Markowetz, Menthal: A Framework for Mobile Data Collection and Analysis, in *Proceedings of the 2016 ACM International Joint Conference on Pervasive and Ubiquitous Computing: Adjunct*. UbiComp '16 (ACM, 2016), pp. 624–629. https://doi.org/10.1145/2968219.2971591
2. J. Asselbergs, J. Ruwaard, M. Ejdys, N. Schrader, M. Sijbrandij, H. Riper, Mobile phone-based unobtrusive ecological momentary assessment of day-to-day mood: an explorative study. J. Med. Internet Res. **18**(3), e72 (2016). https://doi.org/10.2196/jmir.5505

---

[7]We consider DS2 instead of DS1 here, as DS1 contains several users that probably never used or intended to use the app; see *Note on DS1* above.

3. J. Beel, S. Langer, A. Nürnberger, M. Genzmehr, The impact of demographics (age and gender) and other user-characteristics on evaluating recommender systems, in *Research and Advanced Technology for Digital Libraries (TPDL 2013)*, vol. 8092, ed. by T. Aalberg, C. Papatheodorou, M. Dobreva, G. Tsakonas, C.J. Farrugia. LNCS (Springer, 2013), pp. 396–400. https://doi.org/10.1007/978-3-642-40501-3_45

4. F. Beierle, S.C. Matz, M. Allemand, Smartphone sensing in personality science, in *Mobile Sensing in Psychology: Methods and Applications*, ed. by M.R. Mehl, C. Wrzus, M. Eid, G. Harari, U.E. Priemer (Guilford Press, New York, 2021)

5. F. Beierle, V.T. Tran, M. Allemand, P. Neff, W. Schlee, T. Probst, R. Pryss, J. Zimmermann, Context data categories and privacy model for mobile data collection apps, in *Procedia Computer Science. The 15th International Conference on Mobile Systems and Pervasive Computing (MobiSPC)* 134 (2018), pp. 18–25. https://doi.org/10.1016/j.procs.2018.07.139

6. F. Beierle, V.T. Tran, M. Allemand, P. Neff, W. Schlee, T. Probst, R. Pryss, J. Zimmermann, TYDR—track your daily routine. Android app for tracking smartphone sensor and usage data, in *2018 IEEE/ACM 5th International Conference on Mobile Software Engineering and Systems (MOBILESoft)* (ACM, 2018), pp. 72–75. https://doi.org/10.1145/3197231.3197235

7. F. Beierle, V.T. Tran, M. Allemand, P. Neff, W. Schlee, T. Probst, J. Zimmermann, R. Pryss, What data are smartphone users willing to share with researchers? J. Ambient. Intell. Humaniz. Comput. **11**, 2277–2289 (2020). https://doi.org/10.1007/s12652-019-01355-6

8. D. Ben-Zeev, E.A. Scherer, R. Wang, H. Xie, A.T. Campbell, Next-generation psychiatric assessment: using smartphone sensors to monitor behavior and mental health. Psychiatr. Rehabil. J. **38**(3), 218–226 (2015). https://doi.org/10.1037/prj0000130

9. L. Canzian, M. Musolesi, Trajectories of depression: unobtrusive monitoring of depressive states by means of smartphone mobility traces analysis, *Proceedings of the 2015 ACM International Joint Conference on Pervasive and Ubiquitous Computing*. UbiComp '15 (ACM, 2015), pp. 1293–1304. https://doi.org/10.1145/2750858.2805845

10. B. Cao, L. Zheng, C. Zhang, P.S. Yu, A. Piscitello, J. Zulueta, O. Ajilore, K. Ryan, A.D. Leow, DeepMood: modeling mobile phone typing dynamics for mood detection, in *Proceedings of the 23rd ACM SIGKDD International Conference on Knowledge Discovery and Data Mining*. KDD '17 (Association for Computing Machinery, 2017), pp. 747–755. https://doi.org/10.1145/3097983.3098086

11. D. Carneiro, A.P. Pinheiro, P. Novais, Context acquisition in auditory emotional recognition studies. J. Ambient. Intell. Humaniz. Comput. **8**(2), 191–203 (2017). https://doi.org/10.1007/s12652-016-0391-2

12. G. Chittaranjan, J. Blom, D. Gatica-Perez, Who's who with big-five: analyzing and classifying personality traits with smartphones, in *Proceedings of 2011 15th Annual International Symposium on Wearable Computers* (IEEE, 2011), pp. 29–36. https://doi.org/10.1109/ISWC.2011.29

13. G. Chittaranjan, J. Blom, D. Gatica-Perez, Mining large-scale smartphone data for personality studies. Pers. Ubiquit. Comput. **17**(3), 433–450 (2013). https://doi.org/10.1007/s00779-011-0490-1

14. M.J. Chorley, R.M. Whitaker, S.M. Allen, Personality and location-based social networks. Comput. Hum. Behav. **46**(Supplement C), 45–56 (2015). https://doi.org/10.1016/j.chb.2014.12.038

15. D. Danner, B. Rammstedt, M. Bluemke, L. Treiber, S. Berres, C. Soto, O. John. *Die Deutsche Version Des Big Five Inventory 2 (BFI-2)*. Zusammenstellung Sozialwissenschaftlicher Items Und Skalen. 2016. https://doi.org/10.6102/zis247

16. D. Di Matteo, A. Fine, K. Fotinos, J. Rose, M. Katzman, Patient willingness to consent to mobile phone data collection for mental health apps: structured questionnaire. JMIR Mental Health **5**(3) (2018). https://doi.org/10.2196/mental.9539

17. P. Eskes, M. Spruit, S. Brinkkemper, J. Vorstman, M.J. Kas, The sociability score: app-based social profiling from a healthcare perspective. Comput. Hum. Behav. **59**, 39–48 (2016). https://doi.org/10.1016/j.chb.2016.01.024

18. A.A. Farhan, C. Yue, R. Morillo, S. Ware, J. Lu, J. Bi, J. Kamath, A. Russell, A. Bamis, B. Wang, Behavior vs. Introspection: refining prediction of clinical depression via smartphone sensing data, in *2016 IEEE Wireless Health (WH)* (IEEE, 2016), pp. 1–8. https://doi.org/10.1109/WH.2016.7764553

19. D. Ferreira, V. Kostakos, A.K. Dey, AWARE: mobile context instrumentation framework. Front. ICT **2** (2015). https://doi.org/10.3389/fict.2015.00006

20. A.R. Filippo, Innovating in uncertainty: effective compliance and the GDPR. Harv. J. Law Technol. (2018). https://jolt.law.harvard.edu/digest/innovating-in-uncertainty-effective-compliance-and-the-gdpr

21. W. Fleeson, Toward a structure-and process-integrated view of personality: traits as density distributions of states. J. Pers. Soc. Psychol. **80**(6), 1011–1027 (2001). https://doi.org/10.1037/0022-3514.80.6.1011

22. F.M. Götz, S. Stieger, U.-D. Reips, Users of the main smartphone operating systems (iOS, Android) differ only little in personality. PLOS ONE **12**(5), e0176921 (2017). https://doi.org/10.1371/journal.pone.0176921

23. Y. Huang, H. Xiong, K. Leach, Y. Zhang, P. Chow, K. Fua, B.A. Teachman, L.E. Barnes, Assessing social anxiety using GPS trajectories and point-of-interest data, in *Proceedings of the 2016 ACM International Joint Conference on Pervasive and Ubiquitous Computing.* UbiComp '16. Association for Computing Machinery Sept. 2016, pp. 898–903. https://doi.org/10.1145/2971648.2971761

24. G.C.-L. Hung, P.-C. Yang, C.-C. Chang, J.-H. Chiang, Y.-Y. Chen, Predicting negative emotions based on mobile phone usage patterns: an exploratory study. JMIR Res Protocols **5**(3), e160 (2016). https://doi.org/10.2196/resprot.5551

25. K. Jayarajah, R.K. Balan, M. Radhakrishnan, A. Misra, Y. Lee, LiveLabs: building in-situ mobile sensing & behavioural experimentation testbeds, in *Proceedings of the 14th Annual International Conference on Mobile Systems, Applications, and Services.* MobiSys '16 (ACM, 2016), pp. 1–15. https://doi.org/10.1145/2906388.2906400

26. F. Kargl, R.W. van der Heijden, B. Erb, C. Bösch, Privacy in mobile sensing, in *Digital Phenotyping and Mobile Sensing: New Developments in Psychoinformatics*, ed. by H. Baumeister, C. Montag. Studies in Neuroscience, Psychology and Behavioral Economics (Springer, 2019), pp. 3–12. https://doi.org/10.1007/978-3-030-31620-4_1

27. N. Kiukkonen, J. Blom, O. Dousse, D. Gatica-Perez, J. Laurila, Towards rich mobile phone datasets: lausanne data collection campaign, in *Proceedings of ACM International Conference on Pervasive Computing (ICPS)* (2010)

28. S. Langer, J. Beel, The comparability of recommender system evaluations and characteristics of docear's users, in: *Proceedings of Workshop on Recommender Systems Evaluation: Dimensions and Design (REDD) at the 2014 ACM Conference Series on Recommender Systems (RecSys)* (ACM, 2014), pp. 1–6

29. Y. Li, Y. Zhao, S. Ishak, H. Song, N. Wang, N. Yao, An anonymous data reporting strategy with ensuring incentives for mobile crowd-sensing. J. Ambient. Intell. Humaniz. Comput. **9**(6), 2093–2107 (2018). https://doi.org/10.1007/s12652-017-0529-x

30. R. LiKamWa, Y. Liu, N.D. Lane, L. Zhong, MoodScope: Building a mood sensor from smartphone usage patterns, in *Proceeding of the 11th Annual International Conference on Mobile Systems, Applications, and Services.* MobiSys '13 (ACM, 2013), pp. 389–402. https://doi.org/10.1145/2462456.2464449

31. G. López, G. Marín, M. Calderón, Human aspects of ubiquitous computing: a study addressing willingness to use it and privacy issues. J. Ambient. Intell. Humaniz. Comput. **8**(4), 497–511 (2017). https://doi.org/10.1007/s12652-016-0438-4

32. A. Mehrotra, F. Tsapeli, R. Hendley, M. Musolesi, MyTraces: investigating correlation and causation between users' emotional states and mobile phone interaction, in *Proceedings of the ACM on Interactive, Mobile, Wearable and Ubiquitous Technologies*, vol. 1, no. 3 (Sept. 2017), pp. 83:1–83:21. https://doi.org/10.1145/3130948
33. B. Mønsted, A. Mollgaard, J. Mathiesen, Phone-based metric as a predictor for basic personality traits. J. Res. Pers. **74**, 16–22 (2018). https://doi.org/10.1016/j.jrp.2017.12.004
34. C. Montag, H. Baumeister, C. Kannen, R. Sariyska, E.-M. Meßner, M. Brand, Concept, possibilities and pilot-testing of a new smartphone application for the social and life sciences to study human behavior including validation data from personality psychology. J. Multidiscip. Sci. J. **2**(2), 102–115 (2019). https://doi.org/10.3390/j2020008
35. C. Montag, K. Błaszkiewicz, B. Lachmann, I. Andone, R. Sariyska, B. Trendafilov, M. Reuter, A. Markowetz, Correlating personality and actual phone usage. J. Individ. Differ. **35**(3), 158–165 (2014). https://doi.org/10.1027/1614-0001/a000139
36. C. Montag, K. Błaszkiewicz, R. Sariyska, B. Lachmann, I. Andone, B. Trendafilov, M. Eibes, A. Markowetz, Smartphone usage in the 21st century: who is active on whatsapp? BMC Res. Notes **8**(1), 331 (2015). https://doi.org/10.1186/s13104-015-1280-z
37. S. Saeb, M. Zhang, C.J. Karr, S.M. Schueller, M.E. Corden, K.P. Kording, D.C. Mohr, Mobile phone sensor correlates of depressive symptom severity in daily-life behavior: an exploratory study. J. Med. Internet Res. **17**(7), e175 (2015). https://doi.org/10.2196/jmir.4273
38. R. Schoedel, Q. Au, S.T. Völkel, F. Lehmann, D. Becker, M. Bühner, B. Bischl, H. Hussmann, C. Stachl, Digital footprints of sensation seeking. Zeitschrift für Psychologie **226**(4), 232–245 (2018). https://doi.org/10.1027/2151-2604/a000342
39. V.K. Singh, R.R. Agarwal, Cooperative phoneotypes: exploring phone-based behavioral markers of cooperation, in *Proceedings of the 2016 ACM International Joint Conference on Pervasive and Ubiquitous Computing*. UbiComp '16 (ACM, 2016), pp. 646–657. https://doi.org/10.1145/2971648.2971755
40. V.K. Singh, I. Ghosh, Inferring individual social capital automatically via phone logs. Proc. ACM Human-Comput. Interact. **1**CSCW, 95:1–95:12 (2017). https://doi.org/10.1145/3134730
41. C.J. Soto, O.P. John, The next big five inventory (BFI-2): developing and assessing a hierarchical model with 15 facets to enhance bandwidth, fidelity and predictive power. J. Pers. Soc. Psychol. **113**(1), 117–143 (2017). https://doi.org/10.1037/pspp0000096
42. C. Stachl, S. Hilbert, J.-Q. Au, D. Buschek, A. De Luca, B. Bischl, H. Hussmann, M. Bühner, personality traits predict smartphone usage. Eur. J. Personal. **31**(6), 701–722 (2017). https://doi.org/10.1002/per.2113
43. C. Stachl, Q. Au, R. Schoedel, D. Buschek, S. Völkel, T. Schuwerk, M. Oldemeier, T. Ullmann, H. Hussmann, B. Bischl, M. Bühner, *Behavioral Patterns in Smartphone Usage Predict Big Five Personality Traits*. Preprint. PsyArXiv, June 2019. https://doi.org/10.31234/osf.io/ks4vd
44. T. Stütz, T. Kowar M. Kager, M. Tiefengrabner, M. Stuppner, J. Blechert, F.H. Wilhelm, S. Ginzinger, Smartphone based stress prediction, in *User Modeling, Adaptation and Personalization*, ed. by F. Ricci, K. Bontcheva, O. Conlan, S. Lawless, Lecture Notes in Computer Science (Springer International Publishing, 2015), pp. 240–251. https://doi.org/10.1007/978-3-319-20267-9_20
45. R. Wang, F. Chen, Z. Chen, T. Li, G. Harari, S. Tignor, X. Zhou, D. Ben-Zeev, A.T. Campbell, StudentLife: assessing mental health, academic performance and behavioral trends of college students using smartphones, in *Proceedings of the 2014 ACM International Joint Conference on Pervasive and Ubiquitous Computing*. UbiComp '14 (ACM, 2014), pp. 3–14. https://doi.org/10.1145/2632048.2632054
46. W. Wang, G.M. Harari, R. Wang, S.R. Müller, S. Mirjafari, K. Masaba, A.T. Campbell, Sensing behavioral change over time: using within-person variability features from mobile sensing to predict personality traits. Proc. ACM Interact. Mob. Wearable Ubiquitous Technol. **2**(3), 141:1–141:21 (2018). https://doi.org/10.1145/3264951

47. R. Xu, R.M. Frey, E. Fleisch, A. Ilic, Understanding the impact of personality traits on mobile app adoption—insights from a large-scale field study. Comput. Hum. Behav. **62**(Supplement C), 244–256 (2016). https://doi.org/10.1016/j.chb.2016.04.011

48. X. Zhang, W. Li, X. Chen, S. Lu, MoodExplorer: towards compound emotion detection via smartphone sensing. Proc. ACM Interact. Mob. Wearable Ubiquitous Technol. **1**(4), 176:1–176:30 (2018). https://doi.org/10.1145/3161414

49. J. Zimmermann, W.C. Woods, S. Ritter, M. Happel, O. Masuhr, U. Jaeger, C. Spitzer, A.G.C. Wright, Integrating structure and dynamics in personality assessment: first steps toward the development and validation of a personality dynamics diary. Psychol. Assess. **31**(4), 516–531 (2019). https://doi.org/10.1037/pas0000625

Chapter 5
Smartphone Usage Frequency and Duration in Relation to Personality Traits

Using data collected with TYDR, we conduct a psychoinformatical study analyzing the relationship between smartphone usage and personality. This chapter addresses research task A-Task4 (*Research the relationship between smartphone usage frequency, usage session duration, and personality traits of the user.*). While there are previous studies about the relationship between smartphone usage and personality traits of the users, to the best of our knowledge, we are the first to specifically analyze smartphone usage frequencies and duration in relation to personality traits. The results in this chapter have been published in [5].

5.1 Motivation

Smartphones are already omnipresent in everyday life. In the USA, 95% of families with children under 8 years old have a smartphone.[1] The overall number of smartphone users surpasses three billion and is estimated to grow by several hundred millions in the next years.[2] The smartphone is the most frequently used technical device.[3]

Personality traits are often correlated with many aspects of personal as well as professional life (cf. Sect. 3.1). The research field of personality psychology is mostly based on self-assessments and self-reports rather than measurements of actual behavior [23, 4]. Psychoinformatics, the combination of psychological research and computer science (cf. Sect. 1.1 and Sect. 3.3), can help analyze real human behavior with objective measurements of the smartphone. Mobile data

---

[1] https://www.commonsensemedia.org/zero-to-eight-census-infographic.

[2] https://www.statista.com/statistics/330695/number-of-smartphone-users-worldwide/.

[3] https://vertoanalytics.com/chart-of-the-week-which-devices-are-used-most-often/.

© The Author(s), under exclusive license to Springer Nature Switzerland AG 2021
F. Beierle, *Integrating Psychoinformatics with Ubiquitous Social Networking*,
T-Labs Series in Telecommunication Services,
https://doi.org/10.1007/978-3-030-68840-0_5

collection brings the opportunity to conduct research of human behavior in an immediate way. The different context data types (cf. Sect. 4.2) allow for insights into human lives in a direct fashion. In contrast to self-reports, such direct measurements of behavior can be more precise and have higher ecological validity [13, 23].

Based on our context data categorization in Sect. 4.2, in this section, we focus on the *device* category, more specifically, the screen state. Building on existing research in the field of psychoinformatics, we analyze the relationship between personality traits and two dependent variables: screen wakeups (i.e., turning on the smartphone display) and duration of the usage session. We control for different usages depending on the time of the week (weekdays vs. weekends). While existing research has investigated correlations between personality traits and mean duration of daily smartphone usage, to the best of our knowledge, we are the first to disentangle these two features. Our results help psychologists researching smartphone overuse, which is related to, for example, poor sleep quality [26] and weaker work performance [14]. Additionally, researchers and developers of mobile applications can use the insights to better understand their user base.

In the following, we give an overview about previous studies (Sect. 5.2), before detailing the methodology in Sect. 5.3. In Sect. 5.4, we describe the statistical evaluation, and we present the results in Sect. 5.5. We discuss the results in Sect. 5.6.

5.2   Previous Studies

There are several studies investigating the associations between Big Five personality traits and smartphone usage [6, 8, 13, 12, 23, 22]. The authors of [15] report that they could predict extraversion by smartphone usage. Other studies report modest prediction success for extraversion, openness, conscientiousness, and some facets of emotional stability based on a wide range of features derived from smartphone usage [12, 22]. The most common finding is the association of extraversion with increased smartphone usage, through receiving more calls [6], or a higher number of calls and higher use of photography apps [23]. Conscientiousness was found to be associated to a higher usage of work email but to a lower usage of YouTube, fewer voice calls [6], and a low usage of gaming apps [23]. Individuals who score high on agreeableness tended to have more calls in general, while individuals with high emotional stability had higher number of incoming SMS [6]. Women with high scores on openness demonstrated greater usage of video/audio/music [6]. The authors of [25] predicted personality traits based on lists of installed apps, while the authors of [21] predicted sensation-seeking behavior from a variety of smartphone-based features. To the best of our knowledge, we are the first to specifically investigate smartphone wakeups and usage session duration.

5.3 Methodology

Participants We used TYDR to collect data from Android users that voluntarily installed the app from Google Play. We collected data for an 11-month period from 14 October 2018 to 10 September 2019. Overall, 3,634 users installed TYDR during this time period. The timeframe for which a user had TYDR installed depends on the decision of the user, so the time of installation and the usage duration differ between users. For 1,052 users (29%), there is enough data about their phone usage[4] in order to utilize it in our analysis. The average usage session for those users lasted for 3:01 min (median = 0:53, range 2 s to 12 h). Of all 3,634 users, 765 (21%) filled out the Big Five personality traits questionnaire. Our sample—those that fulfilled the conditions of having filled out the Big Five questionnaire and having at least 1,000 app event log entries associated with them—contains 526 users (14% of all users in our database), whose mean age is 34.57 years (SD: 12.85). Out of those 412 users were male (78.3%) and 114 female (21.7%). The mean age of the male participants was 36.00 (SD = 13.04); the mean age of the female participants was 29.39 (SD = 10.71).

Measures Predictor variables were age, gender, weekday/weekend, and the five personality traits measured with BFI-2. Criterion variables were frequency and duration of smartphone usage sessions per day, measured in seconds irrespective of usage type.

BFI-2 The Big Five personality traits were measured with BFI-2 (cf. Sect. 4.4). Cronbach's alpha ($n = 526$) for each trait is as follows: extraversion ($\alpha = 0.82$), agreeableness ($\alpha = 0.7$), conscientiousness ($\alpha = 0.82$), neuroticism ($\alpha = 0.9$), and openness ($\alpha = 0.81$).

Smartphone Usage Sessions We define a smartphone usage session as the time window in which a user was actively using the phone. Each session starts with turning on the display and ends with turning the screen off. We used Android system's app event log to estimate the user's smartphone usage sessions. Each time a user opens or closes an app, the Android system records an event.[5] Given the appropriate permission by the user, TYDR can access these logs and store them in the TYDR database. We implemented a heuristic for estimating actual usage sessions, including background app removal, detection, and imputation of missing events that were not recorded correctly. Based on 40,140,665 app events, our heuristic yielded 1,826,060 usage sessions in total by the 526 users in our sample. On average, users provided data across 48 days. They had on average 72

---

[4]We considered 1,000 app event log entries to be enough to give some insight into the user's phone usage behavior. More details are provided in the next section.

[5]Additionally, other app events are recorded for other types of actions of apps. For the usage session estimation, we only need the events for opening an app to the foreground and closing it.

sessions per day (median = 63, range 1 to 555), lasting for 221 s (3.7 min) on average (median = 164 s (2.7 min), range 2 s to 4,198 s (70 min)).

5.4 Statistical Analyses

We used R (R version 3.6.0 and the packages lme4, sjstats [3]) to conduct all statistical analyses.[6] We used a random-intercept, random-slope multilevel regression analysis to analyze the effect of personality and time (weekday vs. weekend) on daily smartphone usage patterns (number of wakeups per day, mean session duration per day). The multilevel model (MLM) accounts for the nested design of our study with measurement occasions aggregated on a daily level (Level 1) nested within persons (Level 2). We ran a baseline model without any predictors to determine the overall intraclass correlation (ICC, i.e., the relative extent to which dependent variables varied between people). We then ran a model in which weekend vs. weekday was entered at Level 1, and age, gender, extraversion, openness, neuroticism, agreeableness, and conscientiousness were simultaneously entered on Level 2 (Level 2 variables were all grand-mean centered except gender [9]).[7]

The final model is the following:

Level 1 (within person):

$$Wakeups\ per\ day\ [Mean\ session\ duration\ per\ day]_{ti}$$
$$= \pi_{0i} + \pi_{1i} Weekday\ vs.\ Weekend_{ti} + e_{ti}$$

Level 2 (between people):

$$\pi_{0i} = \beta_{00} + \beta_{01}Age_i + \beta_{02}Gender_i + \beta_{03}Extraversion_i + \beta_{04}Aggreableness_i$$
$$+ \beta_{05}Conscientiousness_i + \beta_{06}Neuroticism_i + \beta_{07}Openness_i + r_{01}$$

Level 2:

$$\pi_{1i} = \beta_{10} + r_{1i}$$

We used R^2_{GLMM} [19, 20] as a measure of explained variance, which can be interpreted like the traditional R^2 in regression analyses. $R^2_{marginal}$ represents the proportion of variance explained by the fixed effects alone. As effect size measure, we used standardized β and 95% confidence intervals.

---

[6]The content of this section, previously published in [5], was primarily produced by the coauthors of that paper. The content is included here for completeness' sake.

[7]We have also calculated the models by including the number of days participants had the TYDR app installed as a measure of participant motivation. This predictor was non-significant in all models.

One could argue that having TYDR installed and seeing one's usage pattern visualized may change the user's behavior. In order to account for this, we calculated MLMs with an additional variable at Level 1 reflecting the time since the first usage of TYDR. Because this variable revealed only tiny effects and beta values for the other predictors did not substantially change, we did not include this variable into the final analyses in order to keep the model comprehensible.

5.5 Results

The multilevel analysis revealed several statistically significant predictors of smartphone usage (see Table 5.1), respectively, time of the week (weekdays vs. weekends), age, gender, and the three personality traits extraversion, neuroticism, and conscientiousness.[8,9]

There was a noticeable lower number of screen wakeups during weekends compared to weekdays ($\beta = -0.10$, 95% CI $[-0.11, -0.08]$). On the other hand, mean session duration per day ($\beta = 0.06$, 95% CI $[0.04, 0.07]$) was higher during the weekend compared to the weekdays. That means that reaching for a smartphone was more frequent during weekdays with shorter duration of usage, while during the weekend, screen wakeups were less frequent while the mean duration of sessions per day was longer.

Age was a significant negative predictor for the number of screen wakeups per day ($\beta = -0.19$, 95% CI $[-0.25, -0.13]$) and a significant positive predictor for the mean session duration per day ($\beta = 0.06$, 95% CI $[0.01, 0.12]$). Younger users had a higher number of screen wakeups per day, while older users had higher mean session duration per day. In other words, younger people seem to check their smartphone more often, but with a shorter duration, while older users spend more time using their smartphone per session.

Participants' gender was also a significant predictor for the mean session duration per day ($\beta = -0.06$, 95% CI $[-0.12, -0.01]$). Female participants spent 31.6 s more time per session than males.

Extraversion was a significant predictor for the number of screen wakeups per day ($\beta = 0.11$, 95% CI $[0.04, 0.17]$), meaning that higher extraversion was associated with more frequent smartphone checking. Additionally, neuroticism was also a significant predictor for the same variable ($\beta = 0.12$, 95% CI $[0.05, 0.18]$), which means that higher neuroticism was associated with higher number of phone checking per day. The personality trait conscientiousness was also found to be a significant predictor for the mean session duration per day ($\beta = -0.14$, 95% CI

---

[8]The content of this section, previously published in [5], was primarily produced by the coauthors of that paper. The content is included here for completeness' sake.

[9]For intercorrelations and for possible gender * personality interactions, refer to [5].

Table 5.1 Results of the multilevel analysis

		Fixed						Random	
		Coeff.	β	95% CI	B	SE	t	Coeff.	SD
Screen wakeups per day									
Intercept	β_{00}				72.0	3.14	22.89***	r_{0i}	33.9
Within-person									
Weekend (ref weekday)	β_{10}	-0.10		-0.11 – -0.08	-9.6	0.82	-11.78***	r_{1i}	13.8
Between-person									
Age	β_{01}	-0.19		-0.25 – -0.13	-0.7	0.11	-6.21***		
Gender (ref female)	β_{02}	0.04		-0.02 – 0.10	4.2	3.52	1.20		
Extraversion	β_{03}	0.11		0.04 – 0.17	7.2	2.33	3.08**		
Agreeableness	β_{04}	>-0.01		-0.07 – 0.06	-0.3	2.86	-0.10		
Conscientiousness	β_{05}	0.01		-0.06 – 0.07	0.6	2.28	0.26		
Neuroticism	β_{06}	0.12		0.05 – 0.18	6.6	1.96	3.34***		
Openness	β_{07}	-0.02		-0.08 – 0.05	-1.0	2.16	-0.46		
ICC = 57.5%, $R^2_{marginal}$ = 6.5%									
Mean session duration per day									
Intercept	β_{00}				225.0	12.50	18.01***	r_{0i}	120.0
Within-person									
Weekend (ref weekday)	β_{10}	0.06		0.04 – 0.07	26.1	3.11	8.38***	r_{1i}	37.9
Between-person									
Age	β_{01}	0.06		0.01 – 0.12	1.0	0.45	2.28*		
Gender (ref female)	β_{02}	-0.06		-0.12 – -0.01	-31.6	14.31	-2.21*		
Extraversion	β_{03}	<0.01		-0.06 – 0.06	0.6	9.49	0.06		
Agreeableness	β_{04}	-0.02		-0.07 – 0.04	-6.8	11.65	-0.59		
Conscientiousness	β_{05}	-0.14		-0.20 – -0.08	-41.1	9.28	-4.43***		
Neuroticism	β_{06}	0.01		-0.05 – 0.08	3.8	7.98	0.48		
Openness	β_{07}	0.03		-0.03 – 0.09	8.7	8.81	1.00		
ICC = 39.1%, $R^2_{marginal}$ = 2.8%									

Notes: All Level 2 variables were grand-mean centered except for gender and weekend. *CI* Confidence interval
$p < 0.05$; ** $p < 0.01$; *** $p < 0.001$

$[-0.20, -0.08]$): a high level of conscientiousness was associated with a shorter duration of the mean session per day.

Furthermore, because our criterion measures were non-normally distributed, we recalculated all analyses by using log-transformed measures of screen wakeups per day and mean session duration per day. None of the significances changed except for the mean session duration per day analysis, where the very small effect for gender ($\beta = -0.06$; see Table 5.1) was now not significant anymore ($\beta = -0.05$, p = 0.14). In order to keep the beta values interpretable, we present the results for the non-log-transformed criterion measures.

Additionally, we analyzed a potential self-selection bias (i.e., participants might be different from non-participants) regarding our dependent variables. The sample size of non-participants (i.e., users who did not fill out the BFI-2 personality traits questionnaire) was $n = 523$. We calculated MLMs with only the group variable (participant vs. non-participant) as the predictor. We did not find any significant differences, neither for wakeup frequencies nor for session duration on a day level.

5.6   Discussion

In this study, we investigated how personality traits and demographic properties are associated with smartphone usage. We found that both extraversion and neuroticism were associated with a higher number of screen wakeups per day, and conscientiousness was related to shorter session durations per day.

Our results regarding extraversion are broadly in line with previous studies showing connections of extraversion with higher frequency and duration of calls or other communication behavior [6, 12, 15, 17, 23, 16]. In the study by Chittaranjan et al. [6], emotional stability was connected to a higher number of incoming SMS, while our study showed neuroticism to be related to higher usage of the smartphone measured as the number of wakeups per day. Azucar et al. reported individuals who are highly conscientious are using social media less [2]. Our results showed decreased usage for conscientious users as well, as those individuals tend to have shorter duration of smartphone usage sessions. [18] investigated correlations between gender, age, personality, and usage of WhatsApp. Similar to our findings about general smartphone usage, they reported that female and younger users were using WhatsApp more, while conscientiousness was correlated with a shorter length of WhatsApp usage. Andone et al. reported similar findings about age, gender, and smartphone usage [1]. They reported that the daily mean of phone usage time was higher for female and younger participants but did not investigate wakeup frequencies or correlations with personality.

We can speculate somewhat about the nature of the relationships between neuroticism and extraversion and higher phone usage. For extraversion, the reason for increased usage might be social, e.g., checking the phone for a message that came in [7, 10]. For higher levels of neuroticism, the reason might be anxiousness with respect to missing important things [24]. Longer usage sessions on weekends could indicate that there is more time for longer searches or tasks. We do not know if the smartphone was used by the user only for private purposes, only for business purposes, or for both. More screen wakeups during weekdays could also be a result of work-related phone usage, which is lesser during weekends. Younger people might show a different phone usage pattern because they might demonstrate a different approach to technology in general. Users scoring higher on the trait of conscientiousness may consciously reduce their phone usage to not be distracted [7]. Further studies are required to confirm these interpretations. Analyses of used apps during the usage sessions could also deepen the understanding of the relationship between personality traits and smartphone usage.

When interpreting the results of the present study, several limitations have to be considered. On a technical level, usage session estimation is not as straightforward as it seems. One way to track smartphone usage is to track the display's state. It can only be tracked while TYDR is running. Due to fragmentation—having many different devices with different Android versions and different software adaptations by smartphone manufacturers—and because other apps can interfere with background processes, it is not always possible to ensure that an app is not closed while running in the background. This leads to display state events being missed. Given these constraints, we opted to use the Android system's app event log to determine usage sessions. In contrast to the display state events, Android itself already records the app events. This has the advantage that even when TYDR is not running for some time, the app events can still be tracked. The errors and potentially faulty records that we found in the app event logs were processed by a simple heuristic that we implemented which removed background apps and imputed missing app closing events.

Out of the 3,634 installations, we only had enough data from 526 users that we could analyze for this study. Larger sample sizes could be helpful for confirming our results. In this study, we only analyzed overall smartphone usage. With TYDR, we collected a lot of other data points (cf. Sect. 4.3) that could give more insights about the relationship between smartphone usage and personality traits. While iOS does not allow for such detailed data tracking, we would expect similar results [11]. Although we could make assumptions about the causalities between personality traits and smartphone usage, this study is only correlational. More extensive user studies, e.g., with user interviews, could investigate causalities more deeply but would also make it more difficult to have large sample sizes.

One strength of our study, on the other hand, is that the data was collected in daily life, increasing the ecological validity of the results. Another strength is the rather large sample size, especially compared to other previous studies (cf. Sect. 3.3). Overall, our findings indicate that, for TYDR users, personality traits were associated with smartphone usage.

References

1. I. Andone, K. Błaszkiewicz, M. Eibes, B. Trendafilov, C. Montag, A. Markowetz, How age and gender affect smartphone usage, in *Proceedings of the 2016 ACM International Joint Conference on Pervasive and Ubiquitous Computing: Adjunct.* UbiComp '16 (ACM, 2016), pp. 9–12. https://doi.org/10.1145/2968219.2971451
2. D. Azucar, D. Marengo, M. Settanni, Predicting the big 5 personality traits from digital footprints on social media: a meta-analysis. Personal. Individ. Differ. **124**, 150–159 (2018). https://doi.org/10.1016/j.paid.2017.12.018
3. D. Bates, M. Mächler, B. Bolker, S. Walker, Fitting linear mixed-effects models using Lme4. J. Stat. Softw. **67**(1), 1–48 (2015). https://doi.org/10.18637/jss.v067.i01
4. R.F. Baumeister, K.D. Vohs, D.C. Funder, Psychology as the science of self-reports and finger movements: whatever happened to actual behavior? Perspect. Psychol. Sci. **2**(4), 396–403 (2007). https://doi.org/10.1111/j.1745-6916.2007.00051.x
5. F. Beierle, T. Probst, M. Allemand, J. Zimmermann, R. Pryss, P. Neff, W. Schlee, S. Stieger, S. Budimir, Frequency and duration of daily smartphone usage in relation to personality traits. Digital Psychol. **1**(1), 20–28 (2020). https://doi.org/10.24989/dp.v1i1.1821
6. G. Chittaranjan, J. Blom, D. Gatica-Perez, Mining large-scale smartphone data for personality studies. Pers. Ubiquit. Comput. **17**(3), 433–450 (2013). https://doi.org/10.1007/s00779-011-0490-1
7. P.T. Costa Jr, R.R. McCrae, The revised NEO personality inventory (NEO-PI-R), in *The SAGE Handbook of Personality Theory and Assessment: Volume 2 — Personality Measurement and Testing* (SAGE Publications Ltd, 2008), pp. 179–198. https://doi.org/10.4135/9781849200479
8. Y.-A. de Montjoye, J. Quoidbach, F. Robic, A.S. Pentland, Predicting personality using novel mobile phone-based metrics, in *Social Computing, Behavioral-Cultural Modeling and Prediction* ed. by A.M. Greenberg, W.G. Kennedy, N.D. Bos. Lecture Notes in Computer Science (Springer, 2013), pp. 48–55. https://doi.org/10.1007/978-3-642-37210-0_6
9. C.K. Enders, D. Tofighi, Centering predictor variables in cross-sectional multilevel models: a new look at an old issue. Psychol. Methods **12**(2), 121–138 (2007). https://doi.org/10.1037/1082-989X.12.2.121
10. L.R. Goldberg, J.A. Johnson, H.W. Eber, R. Hogan, M.C. Ashton, C. R. Cloninger, H.G. Gough, The international personality item pool and the future of public-domain personality measures. J. Res. Pers. . Proceedings of the 2005 Meeting of the Association of Research in Personality **40**(1), 84–96 (2006). https://doi.org/10.1016/j.jrp.2005.08.007
11. F.M. Götz, S. Stieger, U.-D. Reips. Users of the main smartphone operating systems (iOS, Android) differ only little in personality. PLOS ONE **12**(5), e0176921 (2017). https://doi.org/10.1371/journal.pone.0176921
12. G.M. Harari, S.R. Müller, C. Stachl, R. Wang, W. Wang, M. Bühner, P.J. Rentfrow, A.T. Campbell, S.D. Gosling, Sensing sociability: individual differences in young adults' conversation, calling, texting, and app use behaviors in daily life. J. Pers. Soc. Psychol. (2019). https://doi.org/10.1037/pspp0000245
13. G.M. Harari, N.D. Lane, R. Wang, B.S. Crosier, A.T. Campbell, S.D. Gosling, Using smartphones to collect behavioral data in psychological science: opportunities, practical considerations, and challenges. Perspect. Psychol. Sci. **11**(6), 838–854 (2016). https://doi.org/10.1177/1745691616650285
14. K. Lanaj, R.E. Johnson, C.M. Barnes, Beginning the workday yet already depleted? Consequences of late-night smartphone use and sleep. Organ. Behav. Hum. Decis. Process. **124**(1), 11–23 (2014). https://doi.org/10.1016/j.obhdp.2014.01.001
15. B. Mønsted, A. Mollgaard, J. Mathiesen, Phone-based metric as a predictor for basic personality traits. J. Res. Pers. **74**, 16–22 (2018). https://doi.org/10.1016/j.jrp.2017.12.004
16. C. Montag, H. Baumeister, C. Kannen, R. Sariyska, E.-M. Meßner, M. Brand, Concept, possibilities and pilot-testing of a new smartphone application for the social and life sciences to study human behavior including validation data from personality psychology. J. Multidiscip. Sci. J. **2**(2), 102–115 (2019). https://doi.org/10.3390/j2020008

17. C. Montag, K. Błaszkiewicz, B. Lachmann, I. Andone, R. Sariyska, B. Trendafilov, M. Reuter, A. Markowetz, Correlating personality and actual phone usage. J. Individ. Differ. **35**(3), 158–165 (2014). https://doi.org/10.1027/1614-0001/a000139
18. C. Montag, K. Błaszkiewicz, R. Sariyska, B. Lachmann, I. Andone, B. Trendafilov, M. Eibes, A. Markowetz, Smartphone usage in the 21st century: who is active on whatsApp? BMC Res. Notes **8**(1), 331 (2015). https://doi.org/10.1186/s13104-015-1280-z
19. S. Nakagawa, P.C.D. Johnson, H. Schielzeth, The coefficient of determination R2 and intra-class correlation coefficient from generalized linear mixed-effects models revisited and expanded. J. R. Soc. Interface **14**(134) (2017). https://doi.org/10.1098/rsif.2017.0213
20. S. Nakagawa, H. Schielzeth, A general and simple method for obtaining R2 from generalized linear mixed-effects models. Methods Ecol. Evol. **4**(2), 133–142 (2013). https://doi.org/10. 1111/j.2041-210x.2012.00261.x
21. R. Schoedel, Q. Au, S.T. Völkel, F. Lehmann, D. Becker, M. Bühner, B. Bischl, H. Hussmann, C. Stachl, Digital footprints of sensation seeking. Zeitschrift für Psychologie **226**(4), 232–245 (2018). https://doi.org/10.1027/2151-2604/a000342
22. C. Stachl, Q. Au, R. Schoedel, D. Buschek, S. Völkel, T. Schuwerk, M. Oldemeier, T. Ullmann, H. Hussmann, B. Bischl, M. Bühner, *Behavioral Patterns in Smartphone Usage Predict Big Five Personality Traits*. Preprint. PsyArXiv, June 2019. https://doi.org/10.31234/osf.io/ks4vd
23. C. Stachl, S. Hilbert, J.-Q. Au, D. Buschek, A. De Luca, B. Bischl, H. Hussmann, M. Bühner, Personality traits predict smartphone usage. Eur. J. Personal. **31**(6), 701–722 (2017). https:// doi.org/10.1002/per.2113
24. H. Stead, P.A. Bibby, Personality fear of missing out and problematic internet use and their relationship to subjective well-being. Comput. Hum. Behav. **76**, 534–540 (2017). https://doi. org/10.1016/j.chb.2017.08.016
25. R. Xu, R.M. Frey, E. Fleisch, A. Ilic, Understanding the impact of personality traits on mobile app adoption—insights from a large-scale field study. Comput. Hum. Behav. **62**(Supplement C), 244–256 (2016). https://doi.org/10.1016/j.chb.2016.04.011
26. S. Yogesh, S. Abha, S. Priyanka, Mobile usage and sleep patterns among medical students. Indian J. Physiol. Pharmacol. **58**(1), 100–103 (2014)

Chapter 6
Interim Conclusions for Part I: Privacy-Aware Mobile Sensing in Psychoinformatics

Psychoinformatics can help deepen the scientific understanding of people and their behavior. Mobile sensing is a tool in achieving that goal by providing objective measurements sensed by sensors and usage statistics. When collecting data via mobile sensing, the privacy of the users has to be considered. We found a lack of structured approaches to privacy in mobile sensing systems for psychoinformatics and developed PM-MoDaC (Privacy Model for Mobile Data Collection Applications). Our study about smartphone usage frequency and duration in relation to personality traits serves as an example study with TYDR while applying PM-MoDaC. The results are important for psychologists, for example, when dealing with smartphone overuse, for mHealth (mobile health) app developers, as well as for researchers and software developers for understanding their user base.

Not only does personality have relationships with smartphone usage but also with the formation of interpersonal relationships. In the following, we introduce our concepts, methods, and prototypes showing how to integrate insights from psychology and social sciences into ubiquitous social networking applications in order to ultimately support and improve real-life social interactions between people. The relationship between smartphone data and personality forms the conceptual level we continue to follow. On a technical level, this again implies the collection of data via mobile sensing. In social networking scenarios, however, instead of one entity collecting the data, we have interactions between users and need to process the data to offer services. A recurring theme is privacy. In social networking scenarios, privacy awareness has to take into account the sharing of data with other users and service providers.

F. Beierle, *Integrating Psychoinformatics with Ubiquitous Social Networking*,
T-Labs Series in Telecommunication Services,
https://doi.org/10.1007/978-3-030-68840-0_6

Part II
Mobile Sensing in Ubiquitous Social Networking

Chapter 7
Overview

This part of this thesis aims at improving social well-being by using findings from psychology and social sciences in social networking services to improve social interactions. We research and develop services based on the principles from Part I: mobile sensing and the relationship between smartphone data and personality. In this part, we thus address research question B, relating to the implications that psychological research results have on the design of social networking applications. We start by extensively reviewing related work in Chap. 8 (B-Task1). We develop a concept for seamless smartphone-mediated incentivization of social interaction between strangers, SimCon, in Chap. 9 (B-Task2). Based on the developed concept, we propose and evaluate metrics for similarity estimation based on probabilistic data structures in Chap. 10 (B-Task3). Both Chaps. 11 and 12 serve as example applications for ubiquitous social networks (B-Task4). Chapter 11 presents a mobile platform for recommender systems, while Chap. 12 presents an application for groups of users, generating music playlist for people currently in proximity.

Chapter 8
Related Work

This chapter addresses research task B-Task1 (*Survey related work in the field of ubiquitous social networking.*). After presenting the fundamentals about the related concepts from psychology and social sciences in Sect. 8.1, we extensively survey related work in the field of ubiquitous social networking in Sect. 8.2. Parts of this chapter have been published in [8, 9].

8.1 Personality, Homophily, Propinquity, and Social Networking

In Part I, we presented the importance of the concept of *personality* with respect to everyday preferences and smartphone usage. The user's personality and other concepts from psychological and social sciences research that we will address in this section are also valuable aspects for social networking scenarios.

Social networking, here meaning the interaction with other people and the forming of social bonds, is of crucial importance for various aspects of life. Scheff even describes the "maintenance of social bonds as the most crucial human motive" [61]. Studies confirm this; Wilcox et al., for example, found a student's social network to be of crucial importance for academic success [74].

We will analyze what implications research from other fields has for the research and design of (mobile) social networking services. Relevant questions that psychological and social sciences research can help answer are, for example, "What factors play a role in the structuring of social networks?" or "Under what conditions do people become friends?", i.e., "When do people create edges in a social graph[1]?"

---

[1]The social graph indicates connections between users in social networking services; more details follow in the next sections.

© The Author(s), under exclusive license to Springer Nature Switzerland AG 2021
F. Beierle, *Integrating Psychoinformatics with Ubiquitous Social Networking*,
T-Labs Series in Telecommunication Services,
https://doi.org/10.1007/978-3-030-68840-0_8

We argue that *homophily* and *propinquity* are two crucial factors that both structure social networks and are deciding factors for the forming of new social bonds. The implication for computer science research is that the similarity of various characteristics or properties of users can be exploited for social networking services.

Homophily[2] is the concept that people tend to associate themselves with other people who are similar to them. According to McPherson et al., this principle structures network ties of every type, including friendship, work, or partnership [47]. Some of the categories in which people have homophilic contacts are ethnicity, age, education, gender, interest, educational background, or consuming habits. McPherson et al. determine similarity as the strongest predictor for friendship. Fischer also describes how friends tend to be similar to each other in aspects like age, educational background, profession, or income [28]. He describes these aspects as signs that might show belief in the same values and attitudes.

Additionally to those demographic and behavioral aspects, personality traits are also determining factors in forming social bonds. Not only does the personality of a person influence a multitude of aspects, e.g., job performance, satisfaction, or romantic success [29], personality is also a "key determinant for the friendship formation process" [13]. In their study, Selfhout et al. show the importance of homophily for friendship networks [63]. For three of the Big Five personality traits—openness to experience, extraversion, and agreeableness—they conclude that people tend to select friends with similar levels of those traits.

In computer-science-related research, these findings are confirmed, for example, in a study by Nass and Lee [52]. They used computer-generated voices that read book reviews to participants of a study. Noticing cues about the personality trait extraversion displayed by the computer voice, the users showed a higher attraction to the voice that matched their own score on the extraversion scale.

In a more recent study from 2018, Parkinson et al. used fMRI (functional magnetic resonance imaging) to scan neural responses when watching naturalistic movies [54]. The closer the study participants were in their social network, the bigger was the similarity of their neural responses, indicating similar perceptions and responses to the world.

Propinquity describes the effect that people tend to like people whom they meet and interact with regularly [46, 27]. The more people meet, the more likely is the forming of a friendship. This is often related to the *mere-exposure effect* [78]. This effect describes how repeated exposure to something increases the likelihood of liking it. In computer-science-related research, this effect has, for example, been observed by Kraut et al. [39]. They showed that mere physical proximity in offices strongly correlated with higher collaborations. Kraut et al. use their finding to propose concepts for computer-supported collaborative work.

The most basic conclusion we draw from the studies described above is that similarity is a key factor for social networking scenarios. This includes similarity in demographic properties, personality traits, neural responses, and physical proximity.

---

[2]Homophily is also sometimes referred to as the *similarity-attraction effect*.

8.2   Ubiquitous Social Networking

In order to define the term *ubiquitous social networking*, we first give definitions of the related terms of *social network(ing) services* (SNSs) and *ubiquitous computing*. After clarifying additional terminology, we survey related work in detail (B-Task1). First, we look into related survey papers, and then we investigate related work from 2005 to 2019, split into intervals of 5 years. We highlight other fields of related work before summarizing our key findings.

Social Network(ing) Services By the definition of Boyd and Ellison, *social network sites* consist of three major components: a profile, connections to other users, and the ability to view and traverse the list of connections [12]. Boyd and Ellison chose the term *network* over *networking* because the latter implies the initiation of contact with yet unknown people, but the former reflects the mapping of an existing social network onto a virtual one. The definition is rather broad. Rohani and Hock, using the term *social networking services*, define such services as having the possibility of connecting with known and unknown people, message exchange, and the sharing of content [57]. Thelwall developed a typology to distinguish three different functions that *social networking sites* can fulfill: *socializing* with existing contacts, *networking* with new contacts, and *social navigation*—navigating specific content via connection with other users, e.g., Goodreads[3] for books [69].

Ubiquitous Computing In his seminal paper published in 1991, Weiser coined the term *ubiquitous computing* [73]. He describes how computer technology could become ubiquitous, it's presence unnoticeable, and user interaction with it seamlessly integrated into other everyday activities. As a comparison, he describes the written word as a technology that is so pervasive that it doesn't need specific attention and can be found anywhere. Weiser envisioned computers recognizing people entering rooms and adapting to their presence. Advances in technology, wireless communication, and the rise of the Internet and specifically the smartphone bring Weiser's visions and predictions into reality. Ubiquitous computing is also referred to as *pervasive computing* [58] or *ambient intelligence*. There are several bordering research fields, for example, mobile computing, distributed systems, context-aware computing, and human-computer interaction.

Ubiquitous Social Networking (USN) combines the two ideas. Looking at the three functions fulfilled by SNSs (cf. above), this includes the interaction with peo-ple already known (*socializing*), (incentivization of) *networking* with yet unknown people, and services for specific interests (related to *social navigation*).

While smartphones were becoming more and more common, *mobile social networking* was often used to refer to the mobile access to an existing SNS like Facebook. The capabilities of the modern smartphone go beyond mere access to web services though and offer opportunities for services that go beyond those offered in a

---

[3]https://www.goodreads.com/.

traditional SNS. USN focuses on users in proximity and on offering services related
to interaction with friends or strangers.

There are several other terms that are used to refer to same basic idea, for
example: ephemeral social network [16], local social network [60], proximity-based
network (PBN) [41], mobile social networking in proximity (MSNP) [15, 71],
proximity mobile social network (PMSN) [33], pervasive social computing [62],
ad hoc social network [32], or spontaneous social network [53].

Our Terminology In the context of SNSs, often, different types of architectures are
discussed. The terms *distributed* and *decentralized* are often used interchangeably.
We want to make a distinction though. For this thesis, centralization refers to the
governance of a system. In a centralized system, one service provider holds control
over the whole system and stores and processes all the data. Distributed, on the
other hand, refers to computation and data storage. For example, Facebook is a
centralized SNS provider that can utilize computing and data storage in a distributed
fashion. Email is an example for a decentralized service: server providers interact
via standardized interfaces, and the user can choose his/her service provider or host
his/her own email server.

Among the most frequent concerns with centralized SNSs is the potential misuse
of data and loss of user privacy. A single service provider holds all the data and
can use it beyond the level to which users intended to share it with the service
provider [25, 18, 10]. A second concern are so-called lock-in effects. Centralized
SNS providers typically strive to retain their users and make it difficult to switch
to other providers. In the case of Facebook, the only way to connect with other
Facebook users is to be on Facebook.

In the context of web and mobile applications, we regard an application as
privacy-aware when the user has the ability to choose what information he/she wants
to share with other users or with service providers.[4] We argue that decentralized
systems without an entity that holds control over the whole system have the
potential to provide higher privacy awareness and help mitigate lock-in effects.
In the field of SNSs, there have been several approaches to overcome lock-
in effects by decentralizing the service to multiple providers. Researchers and
open-source software developers proposed, for example, the federation of several
providers, using standardized formats for interchanging data between different SNS
providers, or utilizing a publicly maintained directory of SNS identifiers for identity
management (see [30] for an extensive overview about this topic).

Another pair of terms that is often used interchangeably is *peer-to-peer* and
device-to-device. Again, we are making a distinction. For this thesis, peer-to-peer
refers to a distributed architecture in which peers share resources. Peer-to-peer
networks are typically virtual overlay networks on top of an existing network
topology. Peer-to-peer is often pointed out in contrast to the client-server model.

---

[4]Note that the key difference of the privacy awareness discussions here, compared to PM-MoDaC
in Sect. 4.5, is that we are dealing with systems that have other users to interact with and a service
provider offering a service instead of a research institute collecting data.

Device-to-device on the other hand refers to the data transfer from one device to another one, typically without needing any additional infrastructure. Sometimes, the term ad hoc communication is used. Two devices directly communicate with each other, typically without going through any access point or base station. This can be contrasted with infrastructure-based modes of communication where a network is used to reach another endpoint.

Survey Papers In their survey paper about mobile social networks, the authors of [38] distinguish between "centralized," "distributed" (in our terminology, this would be *decentralized*), and hybrid architectures. A "centralized" architecture has a server holding all the SNS's data and functionality. A "distributed" (again, *decentralized*) architecture runs on the nodes of its clients, while a "hybrid" one combines the two approaches by holding some of its data and functionality on central servers.

Wang et al. categorize USN approaches and look into different architectures and technologies used in related projects [71]. They highlight privacy as one of the open research questions and develop a proof-of-concept implementation with WiFi Direct.

Hu et al. highlight opportunistic networking as one of the key aspects of future mobile social networks [34]. According to [34], the first aspect of technical challenges and future research directions is "privacy and security."

Ferrag and Maglaras survey privacy preservation mechanisms in ad hoc social networks [26]. They look at different papers from a variety of application scenarios like military, disaster, mobile social applications, vehicular social applications, etc.

In their survey about mobile social networks, Mao et al. consider "proximity-based" as one category [45]. They highlight actual implementations as one of the open topics in the field.

The survey of Ahmed et al. focuses on utilizing social ties between users for tasks like message forwarding, throughput optimizations, or traffic offloading rather than USN applications [1]. As the biggest challenges for future work, they highlight security and privacy.

In a brief survey about multi-device computing, Zhang et al. consider cellular, Bluetooth, WiFi, and NFC for device-to-device computing [79]. Application scenarios they emphasize are the sharing of content, augmented reality, and mobile big data analysis. As one of the challenges, they list security and privacy.

2005–2009[5] In 2005, Eagle and Pentland published about their project in which they used Bluetooth for proximity detection and a central backend server for a USN service [22]. The server contains user profiles and matchmaking preferences and notifies users of users in proximity when a calculated matching threshold is reached.

With P3, Jones and Grandhi describe the idea of building services on top of information about the relationships and metadata of people and places [35]. The authors describe the design of location-based services and the utilization of co-

---

[5]Sometimes, ideas are published across multiple papers. We then take the first publication as the deciding year for choosing the 5-year period.

location/proximity with other users as a basis for recommender systems. The described scenarios range from the incentivization of contact with strangers in proximity to the annotation of physical locations.

WhozThat follows the idea to use Bluetooth to exchange identifiers from existing online social network platforms like Facebook [5]. The phone can then download the publicly available profile, thus incentivizing social interactions by knowing common interests. Another use case based on the proposed system is a context-aware music player that is based on processing available music preferences from Facebook profiles of users in proximity.

SocialFusion [6, 7] takes into account data not only from existing SNSs but also from mobile phones and nearby sensors. The authors envision scenarios like context-aware video screens, music jukeboxes, or mHealth applications. Proto-typically, they show an app called *SocialFlicks* [7]. A mobile component (phone software) shares the user's Facebook identifier. Using that, a stationary component retrieves the users' favorite movies and plays back trailers based on the collected data.

In a similar manner, Davies et al. develop a smart environment application that displays customized content based on the users in proximity recognized via Bluetooth [19]. The idea here is that the Bluetooth device name can be set by the users so that then a screen can display content based on nearby users.

MobiClique is a fully decentralized middleware for mobile social networking that utilizes Bluetooth transfer of data in a peer-to-peer fashion [55]. Device-to-device data exchange is combined with retrieval of data from central servers. The application scenarios are looking for people with common interests and offering interest-based message boards.

In SMILE, the authors utilize real-life encounters to exchange randomly created symmetric keys in a device-to-device manner [44]. Only users who actually met can then read encrypted messages stored on a server.

Table 8.1 shows an overview of the listed papers, including a short summary of the covered topics.

2010–2014 Spiderweb follows the idea to move from merely accessing existing SNSs to using the specific capabilities of mobile phones for social networking applications [59]. A prototypical USN is developed, including device-to-device communication via Bluetooth, and its usability is evaluated. Application scenarios are those of adding contacts, text messaging, or sharing pictures, for example.

With SCOPE, the authors of [43] develop a peer-to-peer-based USN utilizing DHTs (distributed hash tables). Using the WiFi ad hoc mode, they let phones communicate directly and organize the network's data in a DHT. An example application developed offers text/link sharing and telephony.

E-SmallTalker follows the idea to incentivize social interaction with similar users in physical proximity [76, 14]. Device-to-device communication is used to exchange small amounts of data between phones. In order to find common topics of interest, each user's interests are represented by keywords. Matching keywords indicate matching interests and are used to incentivize social interaction.

Table 8.1 Overview of related work published in 2005–2009

Ref.	Year	Topics
[22]	2005	Incentivization of social interaction; proximity detection; central server notifies about matches in proximity
[35]	2005	Incentivization of social interaction; recommender systems; location-based services
[5]	2008	Incentivization of social interaction; context-aware music player; utilizes user data from Facebook
[6, 7]	2009/2010	Context-aware scenarios like video screens
[19]	2009	Public displays showing customized content based on users in proximity
[55]	2009	Middleware building a peer-to-peer network; incentivization of social interaction; interest-based message boards
[44]	2009	Sharing public keys with other users via device-to-device communication; fully encrypted communication

In [16], the authors develop a concept that facilitates the interaction between people in proximity in workplaces. Their prototype allows to view user profiles of users in proximity and to establish connections in a USN. The authors use WiFi for indoor positions and highlight that they focus on building links between users based on encounters (being at the same place at the same time) instead of co-location (having been at the same place, irrespective of time).

Utilizing WiFi and Bluetooth, the authors of E-Shadow use device-to-device technologies to incentivize social interaction [67, 68]. Users can publish user profiles, and the system also contains a method for indicating the physical direction where to find another user.

In [77], Yu et al. develop prototypical applications for USN scenarios. On the client side, a mobile middleware collects social contexts such as proximity information or phone logs. A server component aggregates the data and provides services based on it. Application scenarios include finding the current physical location of friends and sharing of content in classrooms.

With the Musubi project, the authors of [20] set up a system for group communication utilizing a PKI (public key infrastructure). Users are identified by their public key, and a device-to-device scenario is described for key exchanges. Every communication channel between users and between groups of users is encrypted. While all applications based on Musubi work without any server component, a server with a message queue is necessary to distribute all app content (which is message-based) to all users.

With ShAir, Dubois et al. develop a middleware for mobile peer-to-peer scenarios [21]. The authors develop a prototype for file sharing use cases. Using WiFi, devices share content in a device-to-device manner.

In [75], Yang and Hwang propose a mobile recommender system for point-of-interest (POI) recommendations that utilizes data exchanged in a device-to-device manner. They evaluate their approach with a simulation and a user study.

Table 8.2 Overview of related work published in 2010–2014

Ref.	Year	Topics
[59]	2010	Device-to-device USN; focus on usability
[43]	2010	Device-to-device USN based on DHTs
[76, 14]	2010/2013	Incentivization of social interaction; device-to-device communication
[16]	2011	Incentivization of interaction with co-workers in proximity
[67, 68]	2011/2014	Incentivization of social interaction; physically find specific proximate users; device-to-device communication
[77]	2011	Facilitation of social interaction; middleware collects data; server offers services
[20]	2012	Application platform based on a PKI; device-to-device exchange of public keys; secure communication between users and groups of users
[21]	2013	Middleware for peer-to-peer computing using device-to-device communication; file sharing scenario
[75]	2013	Device-to-device communication; POI recommendations
[2]	2014	Spreading of information in a mobile ad hoc network; crisis scenarios
[4]	2014	Application platform; incentivization of social interaction, crowdsourcing, etc.
[70]	2014	Device-to-device communication; text and photo sharing

In [2], the authors develop a prototype that creates mobile ad hoc networks (MANETs) when there is no other communication infrastructure available. The scenario here is that of immediate crises where information has to be spread fast. The goal is to foster communication between an emergency service and the general public.

Atzmueller et al. develop a platform called *Ubicon* that facilitates different scenarios including enhancing social interactions or crowdsourcing noise level or air quality measurements [4]. The focus is on the server component that processes the clients' data. RFID tags are used for determining proximity of users.

AllJoyn is a software framework that allows devices to communicate with other devices in proximity. At the time of writing, the latest release is from 2017.[6] Wang et al. describe the development process with AllJoyn and develop a prototype that allows for text and photo sharing with users in proximity [70].

Table 8.2 shows an overview of the listed papers, including a short summary of the covered topics.

2015–2019 In their project Min-O-Mee, Lokhandwala et al. use AllJoyn to develop an Android prototype for sharing meeting minutes with people in proximity [41]. In the paper, the authors detail an extension to AllJoyn that allows multi-hop data transmission. Kala et al. use the AllJoyn framework for a disaster-scenario network with the use case of transferring files between devices [37]. The focus is also on multi-hop routing.

---

[6]https://github.com/alljoyn/core-alljoyn/releases.

With C3PO, the authors of [11] develop an application framework for USN with opportunistic networking. The use case is publishing journalistic content in crowded sports stadiums. The content is forwarded to nearby devices and spread across the network.

Both [24] and [42] investigate ephemeral social networks in vehicular scenarios, connecting people in traffic. Luan et al. include a privacy-preserving matching algorithm that matches users on common interests without disclosing the interests directly [42]. The authors of [24] focus on profiles with avatars that reflect the user's affect.

In [36], the authors develop an application that enables group communication and location check-ins. The focus is on security and privacy via encryption. A prototype is developed for Android. The scenarios are described for mobile ad hoc networks and focus on interaction with existing friends.

[17] develops methods of representing different contexts efficiently in order to share them in device-to-device scenarios. The paper focuses on context representations and efficient data structures. The evaluation is done via a simulation.

In [72], the authors analyze data from the device-to-device file sharing application Xender,[7] popular in areas with limited network capabilities. The paper focuses on data analysis, and the authors report that they can predict future file sharing activities.

In [56], the authors develop a Java-based middleware that supports device-to-device communication for Android devices. Given example scenarios are video dissemination in crowded places, searching for missing persons, and communication in disaster situations. The authors highlight security and privacy as one of the main aspects for future work.

In [3], the authors present a device-to-device Android application that utilizes WiFi Direct for ad hoc social networking. The application scenario consists of manually entered interests in a user profile that a system then matches with users in proximity.

In [66], the authors present SmartGroup@Net, an application that enables device-to-device communication between users in an outdoor scenario, utilizing Bluetooth Low Energy. The experimental evaluations show energy-efficient transfer of small amounts of data.

Talk2Me, presented in [64], is a prototype that combines augmented reality with social networking. Basic profile information is sent to nearby devices via device-to-device communication. Along with the profile, a so-called *face signature* is sent, allowing the receiving device to use this signature to recognize the sending user when scanning faces with the camera in smartphones or smart glasses.

With Honeypot, the authors of [51] develop a USN prototype for the scenario of incentivizing social interactions between commuters traveling on the same train. The system was tested in an experiment with a mobile app and backend software

---

[7]https://www.xender.com/.

running on a server in a train. The goal is to incentivize social interaction and promote the well-being of commuters.

With Spontaneous Social Network, Navarro et al. form groups of users based on different contexts like location or profile data [53]. The idea is to connect people without them having to know each other. The focus of the paper is on creating such groups and their perceived quality.

FileLinker, a system proposed by Kwan and Greaves, enables smartphones to exchange data in a device-to-device manner, utilizing WiFi Direct and NFC [40]. By using NFC to set up the WiFi Direct connection, the number of necessary user inputs is minimized.

Table 8.3 shows an overview of the listed papers, including a short summary of the covered topics.

Other Related Works   In 2006, Eagle and Pentland investigated the idea of *reality mining*, i.e., sensing social situations with mobile phones and sensors [23]. The authors analyze individual and group behavior and infer relationships based on the collected data.

There are a few papers related to social networking but not strictly USN. The focus is rather the extension of existing SNSs like Facebook. In CenceMe, the idea is to use sensors of phones to infer what people are doing and to publish it to a social networking service [50]. The *Hot in the City* software uses NFC to post locations to SNSs like Facebook by using NFC tags [31]. In [65], the same application is used

Table 8.3 Overview of related work published in 2015–2019

Ref.	Year	Topics
[41]	2015	Device-to-device communication; sharing meeting minutes with coworkers; multi-hop routing
[11]	2015	Publishing journalistic content during bad network conditions; message forwarding
[24, 42]	2015	Vehicular scenarios; profile matching
[36]	2016	Focus on security and privacy; encrypted group communication; interaction with existing contacts
[17]	2016	Different ways of representing contexts; efficiently sharing them in device-to-device scenarios
[72]	2017	Big data analysis about deployed device-to-device file sharing app
[56]	2017	Middleware; device-to-device communication
[3]	2018	Device-to-device communication; finding similar users in proximity
[66]	2018	Energy-efficient device-to-device message transfer; outdoor scenario
[64]	2018	Augmented reality USN; device-to-device communication; face recognition
[51]	2018	USN on a train; incentivization of social interaction
[53]	2018	Connect people who do not know each other; focus on group creation
[37]	2019	Device-to-device communication disaster scenario; transferring files; multi-hop routing
[40]	2019	Device-to-device communication; file sharing scenario

for connecting people in an SNS when two users touch each other's NFC-enabled phones.

There are several papers about routing in opportunistic networks. Some of those works use some ideas related to USN. In their project SANE, the authors of [48, 49] investigate forwarding of messages in opportunistic networks. The developed routing algorithms utilize user profiles containing interests.

There are some works dealing with specific, narrow, topics related to USNs, for example, [60]. The authors define a *Local Social Network* as the combination of SNSs and opportunistic networking. The paper focuses on the users' privacy and proposes the idea to let users create multiple profiles and let them choose when to share which of the profiles.

Summary In the survey papers related to USN, typical topics are different architectural approaches and device-to-device technologies. Privacy is most often described as an open topic for future research [34, 1, 79]. In the surveyed related work, especially in the earlier works, we see the combination of using specific capabilities of mobile devices and existing SNSs. Common topics in related work are device-to-device communication, context-aware computing, recommender systems, and security/encryption of messages. Regarding scenarios for USN, incentivization of social interaction is the most prevalent. Also common are file sharing, crisis/disaster scenarios, workplace, and some group scenarios.

Regarding the incentivization of social interaction, most of the approaches in the reviewed related work utilize manually entered data (cf. [5, 76, 14, 67, 68, 51]). In our approach SimCon, detailed in Chap. 9, we utilize readily available smartphone data, based on our results from Part I. Similarities in smartphone data cannot evaluated as easily as, for example, manually entered lists of interests. In Chap. 10, we introduce metrics for the automatic similarity estimation based on smartphone data. Although several of the reviewed related works describe device-to-device USN prototypes, actual implementations for different mobile operating systems (Android, iOS) on off-the-shelf smartphones are still a challenge [45]. In Chap. 11, we extensively review available technologies and present our design and implementation for ad hoc device-to-device communication. The group scenarios in the reviewed related work mostly focus on communication, for example, text messaging, within groups [20, 36]. In Chap. 12, we present our USN group application GroupMusic that plays back music for groups of currently present users in a highly automated way, thus implementing the vision of ubiquitous computing.

References

1. M. Ahmed, Y. Li, M. Waqas, M. Sheraz, D. Jin, Z. Han, A survey on socially aware device-to-device communications. IEEE Commun. Surv. Tutor. **20**(3), 2169–2197 (2018). https://doi.org/10.1109/COMST.2018.2820069
2. A. Al-Akkad, C. Raffelsberger, A. Boden, L. Ramirez, A. Zimmermann, Tweeting 'When Online Is off'? Opportunistically creating mobile ad-hoc networks in response to disrupted infrastructure, in *ISCRAM* (2014)

3. N. Aneja, S. Gambhir, Profile-based ad hoc social networking using wi-fi direct on the top of android. Res. Article **2018**. https://doi.org/10.1155/2018/9469536
4. M. Atzmueller, M. Becker, M. Kibanov, C. Scholz, S. Doerfel, A. Hotho, B.-E. Macek, F. Mitzlaff, J. Mueller, G. Stumme, Ubicon and its applications for ubiquitous social computing. N. Rev. Hypermedia Multimedia **20**(1), 53–77 (2014). https://doi.org/10.1080/13614568.2013.873488
5. A. Beach, M. Gartrell, S. Akkala, J. Elston, J. Kelley, K. Nishimoto, B. Ray, S. Razgulin, K. Sundaresan, B. Surendar, M. Terada, R. Han, "WhozThat? Evolving an ecosystem for context-aware mobile social networks. IEEE Netw. **22**(4), 50–55 (2008)
6. A. Beach, M. Gartrell, X. Xing, R. Han, SocialFusion: Context-Aware Inference and Recommendation by Fusing Mobile, Sensor, and Social Data; CU-CS-1059-09 (2009)
7. A. Beach, M. Gartrell, X. Xing, R. Han, Q. Lv, S. Mishra, K. Seada, Fusing mobile, sensor, and social data to fully enable context-aware computing, in *Proceedings of the Eleventh Workshop on Mobile Computing Systems & Applications* (ACM, New York, 2010), pp. 60–65
8. F. Beierle, S. Göndör, A. Küpper, Towards a three-tiered social graph in decentralized online social networks, in *Proceedings of the 7th International Workshop on Hot Topics in Planet-Scale mObile Computing and Online Social neTworking (HotPOST)*, June (ACM, New York, 2015), pp. 1–6. https://doi.org/10.1145/2757513.2757517
9. F. Beierle, K. Grunert, S. Göndör, V. Schlüter, Towards psychometrics-based friend recommendations in social networking services, in *2017 IEEE International Conference on AI & Mobile Services (AIMS)*, June (IEEE, New York, 2017), pp. 105–108. https://doi.org/10.1109/AIMS.2017.22
10. A. Bleicher, The anti-Facebook. IEEE Spectrum **48**(6), 54–82 (2011). https://doi.org/10.1109/MSPEC.2011.5779793
11. A. Boutet, S. Frenot, F. Laforest, P. Launay, N. Le Sommer, Y. Maheo, D. Reimert, C3PO: a network and application framework for spontaneous and ephemeral social networks, in *Web Information Systems Engineering—WISE 2015*, ed. by J. Wang, W. Cellary, D. Wang, H. Wang, S.-C. Chen, T. Li, Y. Zhang. Lecture Notes in Computer Science (Springer, New York, 2015), pp. 348–358
12. D.M. Boyd, N.B. Ellison, Social network sites: definition, history and scholarship. J. Comput.-Mediat. Commun. **13**(1), 210–230 (2007). https://doi.org/10.1111/j.1083-6101.2007.00393.x
13. S. Burgess, E. Sanderson, M. Umaña-Aponte, School Ties: An Analysis of Homophily in an Adolescent Friendship Network. CMPO Working Paper Series 11/267. Centre for Market and Public Organisation (2011)
14. A.C. Champion, Z. Yang, B. Zhang, J. Dai, D. Xuan, D. Li, E-SmallTalker: a distributed mobile system for social networking in physical proximity. IEEE Trans. Parall. Distrib. Syst. **24**(8), 535–1545 (2013). https://doi.org/10.1109/TPDS.2012.251
15. C. Chang, S.N. Srirama, S. Ling, An adaptive mediation framework for mobile P2P social content sharing, in *Service-Oriented Computing*, ed. by C. Liu, H. Ludwig, F. Toumani, Q. Yu. Lecture Notes in Computer Science, vol. 7636 (Springer, Berlin, Heidelberg, 2012), pp. 374–388
16. A. Chin, H. Wang, B. Xu, K. Zhang, H. Wang, L. Zhu, Connecting people in the workplace through ephemeral social networks, in *2011 IEEE Third International Conference on Privacy, Security, Risk and Trust and 2011 IEEE Third International Conference on Social Computing*, October 2011, pp. 527–530. https://doi.org/10.1109/PASSAT/SocialCom.2011.88
17. S. Cho, C. Julien, Chitchat: navigating tradeoffs in device-to-device context sharing. In: *2016 IEEE International Conference on Pervasive Computing and Communications (PerCom)*, March (IEEE, New York, 2016), pp. 1–10. https://doi.org/10.1109/PERCOM.2016.7456512
18. A. Datta, S. Buchegger, L.-H. Vu, T. Strufe, K. Rzadca, Decentralized online social networks, in *Handbook of Social Network Technologies and Applications* (Springer, New York, 2010), pp. 349–378

19. N. Davies, A. Friday, P. Newman, S. Rutlidge, O. Storz, Using bluetooth device names to support interaction in smart environments, in *Proceedings of the 7th International Conference on Mobile Systems, Applications, and Services*. MobiSys '09 (ACM, New York, 2009), pp. 151–164. https://doi.org/10.1145/1555816.1555832
20. B. Dodson, I. Vo, T. Purtell, A. Cannon, M. Lam, Musubi: disintermediated interactive social feeds for mobile devices. in *Proceedings of the 21st International Conference on World Wide Web*, WWW '12 (ACM, New York, 2012), pp. 211–220. https://doi.org/10.1145/2187836.2187866
21. D.J. Dubois, Y. Bando, K. Watanabe, H. Holtzman, ShAir: extensible middleware for mobile peer-to-peer resource sharing, in *Proceedings of the 2013 9th Joint Meeting on Foundations of Software Engineering*. ESEC/FSE 2013 (ACM, New York, 2013), pp. 687–690. https://doi.org/10.1145/2491411.2494573
22. N. Eagle, A. Pentland, Social serendipity: mobilizing social software. Pervas. Comput. IEEE **4**(2), 28–34 (2005)
23. N. Eagle, A.S. Pentland, Reality mining: sensing complex social systems. Person. Ubiquit. Comput. **10**(4), 255–268 (2006) https://doi.org/10.1007/s00779-005-0046-3
24. R. Esmaeilyfard, F. Hendessi, An affective ephemeral social network for vehicular scenarios. J. Amb. Intell. Smart Env. **7**(1), 21–36 (2015)
25. M. Falch, A. Henten, R. Tadayoni, I. Windekilde, Business models in social networking, in *CMI International Conference on Social Networking and Communities* (2009)
26. M.A. Ferrag, L. Maglaras, A. Ahmim. Privacy-preserving schemes for ad hoc social networks: a survey. IEEE Commun. Surv. Tutor. **19**(4), 3015–3045 (2017). https://doi.org/10.1109/COMST.2017.2718178
27. L. Festinger, S. Schachter, K. Back, The spatial ecology of group formation, in *Social Pressures in Informal Groups* (Harper, New York, 1950), pp. 141–161
28. C.S. Fischer, *To Dwell Among Friends: Personal Networks in Town and City* (University of Chicago Press, Chicago, 1982)
29. J. Golbeck, C. Robles, K. Turner, Predicting personality with social media, in *Proceedings of the CHI '11 Extended Abstracts on Human Factors in Computing Systems (CHI EA'11)*. CHI EA '11, May 2011 (ACM, New York, 2011), pp. 253–262. https://doi.org/10.1145/1979742.1979614
30. S.J. Göndör, Seamless Interoperability and Data Portability in the Social Web for Facilitating an Open and Heterogeneous Online Social Network Federation. PhD thesis. Technische Universität Berlin, 2018
31. J. Haikio, T. Tuikka, E. Siira, V. Tormanen, Would you be my friend? Creating a mobile friend network with 'Hot in the City', in *2010 43rd Hawaii International Conference on System Sciences*, January 2010, pp. 1–10. https://doi.org/10.1109/HICSS.2010.2
32. F. Hao, S. Li, G. Min, H.-C. Kim, S.S. Yau, L.T. Yang, An efficient approach to generating location-sensitive recommendations in ad-hoc social network environments. IEEE Trans. Serv. Comput. **8**(3), 520–533 (2015). https://doi.org/10.1109/TSC.2015.2401833
33. W. He, Y. Huang, K. Nahrstedt, B. Wu, Message propagation in ad-hoc-based proximity mobile social networks, in *2010 8th IEEE International Conference on Pervasive Computing and Communications Workshops (PERCOM Workshops)*, March. 2010, pp. 141–146. https://doi.org/10.1109/PERCOMW.2010.5470617
34. X. Hu, T.H.S. Chu, V.C.M. Leung, E.C.-H. Ngai, P. Kruchten, H.C.B. Chan, A survey on mobile social networks: applications, platforms, system architectures, and future research directions. IEEE Commun. Surv. Tutor. **17**(3), 1557–1581 (2015). https://doi.org/10.1109/COMST.2014.2371813
35. Q. Jones, S. Grandhi, P3 systems: putting the place back into social networks. IEEE Intern. Comput. **9**(5), 38–46 (2005). https://doi.org/10.1109/MIC.2005.105
36. J. Joy, E. Chung, Z. Yuan, J. Li, L. Zou, M. Gerla, DiscoverFriends: secure social network communication in mobile ad hoc networks. Wirel. Commun. Mobile Comput. **16**(11), 1401–1413 (2016). https://doi.org/10.1002/wcm.2708

37. S.M. Kala, V. Sathya, S.S. Magdum, T.V.K. Buyakar, H. Lokhandwala, B.R. Tamma, Designing infrastructure-less disaster networks by leveraging the AllJoyn framework, in *Proceedings of the 20th International Conference on Distributed Computing and Networking*. ICDCN '19 (ACM, New York, 2019), pp. 417–420. https://doi.org/10.1145/3288599.3295596

38. N. Kayastha, D. Niyato, P. Wang, E. Hossain, Applications, architectures, and protocol design issues for mobile social networks: a survey. Proc. IEEE **99**(12), 2130–2158 (2011). https://doi.org/10.1109/JPROC.2011.2169033

39. R. Kraut, C. Egido, J. Galegher, Patterns of contact and communication in scientific research collaboration, in *Proceedings of the 1988 ACM Conference on Computer-Supported Cooperative Work CSCW '88* (ACM, New York, 1988), pp. 1–12. https://doi.org/10.1145/62266.62267

40. K. Kwan, B. Greaves, FileLinker: simple peer-to-peer file sharing using Wi-Fi direct and NFC, in *2019 IST-Africa Week Conference (IST-Africa)*, May 2019, pp. 1–9. https://doi.org/10.23919/ISTAFRICA.2019.8764840

41. H. Lokhandwala, S.M. Kala, B.R. Tamma, Min-O-Mee: a proximity based network application leveraging the AllJoyn framework, in *2015 International Conference on Computing and Network Communications (CoCoNet)*, December 2015, pp. 613–619. https://doi.org/10.1109/CoCoNet.2015.7411252

42. T.H. Luan, R. Lu, X. Shen, F. Bai, Social on the road: enabling secure and efficient social networking on highways. IEEE Wirel. Commun. **22**(1), 44–51 (2015). https://doi.org/10.1109/MWC.2015.7054718

43. M. Mani, A.-M. Nguyen, N. Crespi, SCOPE: a prototype for spontaneous P2P social networking, in *2010 8th IEEE International Conference on Pervasive Computing and Communications Workshops (PERCOM Workshops)*, March 2010, pp. 220–225. https://doi.org/10.1109/PERCOMW.2010.5470664

44. J. Manweiler, R. Scudellari, L.P. Cox, SMILE: encounter-based trust for mobile social services, in *Proceedings of the 16th ACM Conference on Computer and Communications Security* (ACM, New York, 2009), pp. 246–255

45. Z. Mao, Y. Jiang, G. Min, S. Leng, X. Jin, K. Yang, Mobile social networks: design requirements, architecture, and state-of-the-art technology. Comput. Commun. **100**(Supplement C), 1–19 (2017). https://doi.org/10.1016/j.comcom.2016.11.006

46. D.M. Marvin, Occupational propinquity as a factor in marriage selection. Quart. Publ. Am. Stat. Assoc. **16**(123), 131–150 (918). https://doi.org/10.1080/15225445.1918.10503750

47. M. McPherson, L. Smith-Lovin, J.M. Cook, Birds of a feather: homophily in social networks. Annu. Rev. Sociol. **27**, 415–444 (2001)

48. A. Mei, G. Morabito, P. Santi, J. Stefa, Social-aware stateless forwarding in pocket switched networks, in *2011 Proceedings IEEE INFOCOM*, April 2011, pp. 251–255. https://doi.org/10.1109/INFCOM.2011.5935076

49. A. Mei, G. Morabito, P. Santi, J. Stefa, Social-aware stateless routing in pocket switched networks. IEEE Trans. Parall. Distrib. Syst. **26**(1), 252–261 (2015). https://doi.org/10.1109/TPDS.2014.2307857

50. E. Miluzzo, N.D. Lane, K. Fodor, R. Peterson, H. Lu, M. Musolesi, S.B. Eisenman, X. Zheng, A.T. Campbell, Sensing meets mobile social networks: the design, implementation and evaluation of the CenceMe application, in *Proceedings of the 6th ACM Conference on Embedded Network Sensor Systems*. SenSys '08 (ACM, New York, 2008), pp. 337–350. https://doi.org/10.1145/1460412.1460445

51. J. Müller, C. Anneser, M. Sandstede, L. Rieger, A. Alhomssi, F. Schwarzmeier, B. Bittner, I. Aslan, E. André, Honeypot: a socializing app to promote train commuters' wellbeing, in *Proceedings of the 17th International Conference on Mobile and Ubiquitous Multimedia*. MUM 2018 (ACM, New York, 2018), pp. 103–108. https://doi.org/10.1145/3282894.3282901

52. C. Nass, K.M. Lee, Does computer-generated speech manifest personality? An experimental test of similarity-attraction, in *Proceedings of the SIGCHI Conference on Human Factors in Computing Systems*. CHI '00, April (Association for Computing Machinery, New York, 2000), pp. 329–336. https://doi.org/10.1145/332040.332452

53. N.D.A.B. Navarro, C.A. da Costa, J.L.V. Barbosa, R.D.R. Righi, Spontaneous social network: toward dynamic virtual communities based on context-aware computing. Exp. Syst. Appl. **95**, 72–87 (2018). https://doi.org/10.1016/j.eswa.2017.11.017

54. C. Parkinson, A.M. Kleinbaum, T. Wheatley, Similar neural responses predict friendship. Nat. Commun. **9**(1), 1–14 (2018). https://doi.org/10.1038/s41467-017-02722-7

55. A.-K. Pietiläinen, E. Oliver, J. LeBrun, G. Varghese, C. Diot, MobiClique: middleware for mobile social networking, in *Proceedings of the 2Nd ACM Workshop on Online Social Networks*. WOSN '09 (ACM, New York, 2009), pp. 49–54. https://doi.org/10.1145/1592665.1592678

56. J. Rodrigues, E.R.B. Marques, L.M.B. Lopes, F. Silva, Towards a middleware for mobile edge-cloud applications, in *Proceedings of the 2nd Workshop on Middleware for Edge Clouds & Cloudlets*. MECC '17 (ACM, New York, 2017), pp. 1:1–1:6. https://doi.org/10.1145/3152360.3152361

57. V.A. Rohani, O.S. Hock, On social network web sites: definition, features, architectures and analysis tools. J. Comput. Eng. **1**, 3–11 (2009)

58. D. Saha, A. Mukherjee, Pervasive computing: a paradigm for the 21st century. Computer **36**(3), 25–31 (2003). https://doi.org/10.1109/MC.2003.1185214

59. A. Sapuppo, Spiderweb: a social mobile network, in *2010 European Wireless Conference (EW)*, April 2010, pp. 475–481. https://doi.org/10.1109/EW.2010.5483495

60. A. Sapuppo, T. Sørensen, Local social networks. in *Computer Communication and Management: International Proceedings of Computer Science and Information Technology*, vol. 5 (IACSIT Press, Singapore, 2011), pp. 15–22

61. T.J. Scheff, *Microsociology: Discourse, Emotion, and Social Structure* (University of Chicago Press, Chicago, 1990)

62. D. Schuster, A. Rosi, M. Mamei, T. Springer, M. Endler, F. Zambonelli, Pervasive social context: taxonomy and survey. ACM Trans. Intell. Syst. Technol. **4**(3), 46:1–46:22 (2013). https://doi.org/10.1145/2483669.2483679

63. M. Selfhout, W. Burk, S. Branje, J. Denissen, M. Van Aken, W. Meeus, Emerging Late adolescent friendship networks and big five personality traits: a social network approach. J. Personal. **78**(2), 509–538 (2010). https://doi.org/10.1111/j.1467-6494.2010.00625.x

64. J. Shu, S. Kosta, R. Zheng, P. Hui, Talk2Me: a framework for device-to-device augmented reality social network, in *2018 IEEE International Conference on Pervasive Computing and Communications (PerCom)*, March (IEEE, New York, 2018), pp. 1–10. https://doi.org/10.1109/PERCOM.2018.8444578

65. E. Siira, V. Törmänen, The impact of NFC on multimodal social media application, in *2010 Second International Workshop on Near Field Communication*, April 2010, pp. 51–56. https://doi.org/10.1109/NFC.2010.16

66. A. Sikora, M. Krzysztoñ, M. Marks, Application of bluetooth low energy protocol for communication in mobile networks, in *2018 International Conference on Military Communications and Information Systems (ICMCIS)*, May 2018, pp. 1–6. https://doi.org/10.1109/ICMCIS.2018.8398689

67. J. Teng, B. Zhang, X. Li, X. Bai, D. Xuan, E-Shadow: lubricating social interaction using mobile phones, in *2011 31st International Conference on Distributed Computing Systems*, June 2011, pp. 909–918. https://doi.org/10.1109/ICDCS.2011.48

68. J. Teng, B. Zhang, X. Li, X. Bai, D. Xuan, E-Shadow: lubricating social interaction using mobile phones. IEEE Trans. Comput. **63**(6) 1422–1433 (2014). https://doi.org/10.1109/TC.2012.290

69. M. Thelwall, Social network sites: users and uses, in *Advances in Computers*, chapter 2 , vol. 76. Social Networking and the Web. Elsevier, January 2009, pp. 19–73. https://doi.org/10.1016/S0065-2458(09)01002-X

70. Y. Wang, L. Wei, Q. Jin, J. Ma, AllJoyn based direct proximity service development: overview and prototype, in *2014 IEEE 17th International Conference on Computational Science and Engineering*, December 2014, pp. 634–641. https://doi.org/10.1109/CSE.2014.138

71. Y. Wang, A.V. Vasilakos, Q. Jin, J. Ma, Survey on mobile social networking in proximity (MSNP): approaches, challenges and architecture. Wirel. Netw. **20**(6), 1295–1311 (2014). https://doi.org/10.1007/s11276-013-0677-7

72. X. Wang, H. Wang, K. Li, S. Yang, T. Jiang, Serendipity of sharing: large-scale measurement and analytics for device-to-device (D2D) content sharing in mobile social networks, in *2017 14th Annual IEEE International Conference on Sensing, Communication, and Networking (SECON)*, June 2017, pp. 1–9. https://doi.org/10.1109/SAHCN.2017.7964925

73. M. Weiser, The computer for the 21st century. Sci. Am. **265**(3), 94–104 (1991). https://doi.org/10.1038/scientificamerican0991-94

74. P. Wilcox, S. Winn, M. Fyvie-Gauld, 'It was nothing to do with the university it was just the people': the role of social support in the first-year experience of higher education. Stud. High. Educ. **30**(6), 707–722 (2005). https://doi.org/10.1080/03075070500340036

75. W.-S. Yang, S.-Y. Hwang, iTravel: a recommender system in mobile peer-to-peer environment. J. Syst. Softw. **86**(1), 12–20 (2013). https://doi.org/10.1016/j.jss.2012.06.041

76. Z. Yang, B. Zhang, J. Dai, A. Champion, D. Xuan, D. Li, E-SmallTalker: a distributed mobile system for social networking in physical proximity, in *2010 IEEE 30th International Conference on Distributed Computing Systems (ICDCS)*, June (IEEE, New York, 2010), pp. 468–477. https://doi.org/10.1109/ICDCS.2010.56

77. Z. Yu, Y. Liang, B. Xu, Y. Yang, B. Guo, Towards a smart campus with mobile social networking, in *2011 International Conference on Internet of Things and 4th International Conference on Cyber, Physical and Social Computing*, October 2011, pp. 162–169. https://doi.org/10.1109/iThings/CPSCom.2011.55

78. R.B. Zajonc, Attitudinal effects of mere exposure. J. Personal. Soc. Psychol. **9**(2), Pt.2, 1–27 (1968). https://doi.org/10.1037/h0025848

79. W. Zhang, H. Flores, P. Hui, Towards collaborative multi-device computing, in *2018 IEEE International Conference on Pervasive Computing and Communications Workshops (PerCom Workshops)*, March (IEEE, New York, 2018), pp. 22–27. https://doi.org/10.1109/PERCOMW.2018.8480262

Chapter 9
SimCon: A Concept for Contact Recommendations

In this chapter, we develop a concept—SimCon—for the most common scenario in USN (ubiquitous social networking), the incentivization of social interaction. This chapter addresses research task B-Task2. While existing approaches typically rely on manually entered profile data, SimCon relies on mobile sensing and can be implemented in a fully automated way, seamless and unobtrusive. An earlier version of this chapter has been published in [5].

9.1 Motivation

SNSs (social networking services) are one of the most used services on the World Wide Web [17]. Two typical elements of a SNS are the *social profile*; containing information about a user, for example his/her interests; and the *social graph*, containing information about the connections between users. The smartphone is the optimal social networking device: it typically has only one user, and looking back at the first part of this thesis, with recent developments in smartphone sensor technologies and available APIs, more and more personal data is available. This data includes, for example, location traces, most frequently used apps, etc., and could potentially extend existing social profiles.

As we have seen in the related work, Sect. 8.2, one of the typical applications in SNSs and USNs is the incentivization of social interaction or, phrased in other terms, the recommendation of new contacts. In traditional, centralized SNSs, typically, the social graph is utilized for this task [44]. Doing so enables the incorporation of graph-based properties like the number of mutual friends. There are also studies that look into the similarity of attributes of neighboring nodes, thus incorporating the social profile in the recommendation process [29].

In USN scenarios, there is usually no social graph of existing connections. Most often, the USN scenarios of incentivizing social interactions are explicitly about two

F. Beierle, *Integrating Psychoinformatics with Ubiquitous Social Networking*,
T-Labs Series in Telecommunication Services,
https://doi.org/10.1007/978-3-030-68840-0_9

strangers meeting, without having any knowledge about the connections between users. Following the results of the first part of the thesis about the relationship between smartphone data and personality, we develop a contact recommendation approach for USN scenarios, called SimCon, based on smartphone data, which can be applied to complete strangers and which does not rely on a social graph.

In 2016, the company *Cambridge Analytica* was in the media because of their alleged success in utilizing psychometrics in targeted political campaign advertisements,[1] though their impact on the campaign remains somewhat unclear [11]. Although the use case is different—targeted advertising instead of contact recommendation—this shows the potential of applying psychological research results to other fields.

This chapter is structured as follows: based on our findings from Chap. 8, we analyze the background regarding offline and online social networking and social profiles (Sect. 9.2). This includes a thorough analysis of the theoretical background of psychometrics and other research from psychology and social sciences in relation to SNSs and mobile devices. We integrate our findings and develop our concept SimCon for contact recommendations for USN scenarios (Sect. 9.3). Next, we set our concept in the context of other contact recommendation approaches (Sect. 9.4).

9.2   Fundamentals and Background

In Sect. 8.1, we gave details about characteristics of social networks from a psychology and social science perspective. These insights give us a fundamental understanding about how and when people form interpersonal bonds in "offline" social networks. In order to ensure that the same concepts hold true in SNSs, we look into existing research on SNSs, personality, and social ties. We propose to use user data instead of an existing social graph for contact recommendations in USN scenarios. In SNSs, user data is typically contained in the social profile. We review existing definitions and components of social profiles and set our findings in relation to smartphone data.

Offline vs. Online Social Networking (Services) Several studies suggest that the findings about (offline) social networks are also valid when dealing with SNSs. Liu claims the social profile is a "performance" by the user who expresses himself/herself by crafting the profile [22]. While this might be true, various studies show that this does not imply that this "performance" distorts the personality that is expressed in the profile. For example, Back et al. conclude in their study that "Facebook Profiles Reflect Actual Personality, Not Self-Idealization," as the title of their paper indicates [2]. In their study, Goldbeck et al. show that Facebook profiles can be used to predict personality [14]. Another study comes to the same conclusion

---

[1]http://www.thetimes.co.uk/article/trump-calls-in-brexit-experts-to-target-voters-pf0hwcts9.

and shows "that Facebook-based personality impressions show some consensus for all Big Five dimensions" [16]. Kosinski et al. show that Facebook *likes* can be used to accurately predict various sensitive user attributes like personality traits, sexual orientation, religious and political views, or use of addictive substances [20].

Not only can we find significant relationships between smartphone data and the user's personality, we can also find further significant relationships between smartphone data and user aspects that are useful for contact recommendations.

Nguyen et al. estimated the similarity of SNS users based on their behavior, e.g., posting, liking, and commenting [31]. The ground truth was a group of volunteers judging the similarity of Twitter users. The general concept observed is that similar people display similar behavior—in this case in Twitter.

There are several papers looking into existing friendships. Njoo et al. show that co-location data can be used to distinguish friends and strangers [32]. The findings show that similar people display similar behavior in the sense of being at the same places more frequently than strangers. Sapiezynski et al. report that while virtual communication like Facebook interactions and phone calls is indicative of stronger interpersonal bonds, it is personal proximity that is even more indicative [37]. This study confirms the finding of friends spending time in physical proximity. By tracking smartphone data, for example, with TYDR, potentially, both aspects, co-location and virtual communication like phone call statistics, can be tracked. In a paper about geolocation traces and data privacy, Gambs et al. highlight that "there is a strong interplay between the geolocated data of an individual and its social network in the sense that knowledge about one can help infer new information about the other" [13].

In more general terms, not only looking at existing friendships, Wang et al. investigate (co-)location traces in [41]. They conclude that "[h]uman mobility could indeed serve as a good predictor for the formation of new links, yielding comparable predictive power to traditional network-based measures."

Social Profiles The social profile is one of the central elements of SNSs. In Boyd and Ellison's definition of *social network sites*, the "public or semi-public profile within a bounded system" is the first defining element and the "backbone" of the SNS [6]. Typical elements of a social profile are "age, location, interests, and an 'about me' section" and a photo. In [35], the social profile is the first defining functionality of an SNS. Here, the authors call the functionality "identity management," as the profile is a "representation of the own person." In a survey paper about SNSs, the social profile is described as the "core" of an online social network [19]. Another recent survey describes the creation and maintenance of user profiles as the "basic functionality" of SNSs [33]. In [15], several SNSs from different categories, like general (e.g., Facebook), business-oriented (e.g., LinkedIn), or special purpose (e.g., Twitter), are analyzed. The social profile is an element that is present in all of those SNSs. Rohani and Hock state that the type of information included in social profiles differs between different SNSs [36]. In their analysis of publicly disclosed Facebook profile information, Farahbakhsh et al. distinguish between personal and interest-based attributes [12]. Personal attributes include a friend list, current city, hometown, gender, birthday, employers, college, and high school. Interest-based attributes are music, movie, book, television, games, teams, sports, athletes, activities, interests, and inspirations. Lampe et al. distinguish between three different types of information: referents, interests, and contact [21]. Referents include verifiable attributes: hometown, high school, residence, and concentration. Contact information are also verifiable, for example, website, email, address, or birthday. As the authors indicate, interests are less verifiable. Interests include an "about me" section, favorite music, movies, TV shows, books, quotes, and political views.

Bao et al. suggest that more detailed user data additional to the data typically available in a social profile can help improve recommendations in SNSs [3]. They suggest that, for example, location histories can serve as indications of shared preferences and interests. We argue that this, in turn, also means that some of the features typically displayed in a social profile can already be determined automatically. If a user frequents a specific restaurant, it could give the same indication as liking its Facebook page. Musical preference is tracked by tracking the music that was played back; favorite POIs or other places are part of the location history. App usage most likely indicates interests as well, e.g., interest in playing specific games or not playing any games. Other features based on smartphone data might go beyond what a typical social profile typically contains, like notification metadata or app traffic.

9.3   SimCon Concept

For our concept SimCon, we combine the findings presented from related work in psychology, social sciences, and computer science. First, we summarize our two key findings that our concept is based on:

- *Similarity is of key importance.* Research in psychology and social sciences indicates that homophily in demographic aspects and personality traits and interests are the best predictors for friendship, i.e., a successful contact recommendation. This aspect, as well as propinquity, i.e., co-location, translates to the importance of user similarity. These findings hold true for offline as well as online social networking.
- *Smartphone data reflects various aspects where we seek similarity.* Both the first part of this thesis and the related work in the previous section highlight the ties between smartphone data and user characteristics like personality traits and interests. Thus, collecting smartphone data can be an unobtrusive way to approximate user characteristics without having to have the user tediously enter information manually.

Combining those insights, in order to make meaningful recommendations for new connections in a USN scenario, we can recommend users who show a similar behavioral pattern with their smartphones.

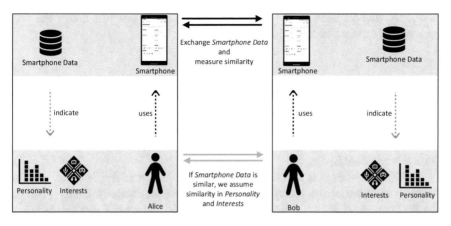

Fig. 9.1 SimCon concept for contact recommendations in USN scenarios

Even without having to know the exact relationships between smartphone data and, e.g., personality traits or preferences and interests, we propose to compare a profile of collected data to receive an indication of similarity in relevant user characteristics.

The SimCon concept works as follows:

1. The smartphone tracks data about its user via mobile sensing.
2. There are relationships between smartphone data and aspects that are relevant for contact recommendations (i.e., interests, personality traits).
3. Similarity in smartphone data indicates similarity in relevant aspects—similar users are recommended as new contacts.

Figure 9.1 visualizes the core idea: smartphone data indicate personality and interests of the user. In order to determine if Alice and Bob are similar to each other, we measure the similarity of their smartphone data.

This approach works with smartphone data only. This means that we do not rely on the social graph and our concept can be applied to complete strangers. Furthermore, by using smartphone data instead of a manually entered social profile, SimCon can be executed completely autonomously: smartphone data can be collected automatically and exchanged automatically.[2]

In the related work given in Sect. 8.2, the social profile often consists of interests that are entered manually (cf. [4, 43, 7, 38, 39, 30]). When reviewing such a profile of a user in proximity, the evaluation of the profile often has to be done manually as well, i.e., the user has to read the profiles of other users in order to decide if she/he finds the other user interesting. In contrast, with SimCon, by estimating the similarity of two profiles, no manual evaluation, i.e., reading of a profile, is

---

[2]Details about data exchange without user interaction follow in Chap. 11.

necessary (details about this follow in Chap. 10.). We can just notify the user whenever a certain similarity threshold is reached.

9.4 Other Contact Recommendation Approaches

In the following, we give details about related work that is specific to contact recommendations in USN scenarios, as well as the common approach of predicting links in traditional SNSs.

In 2005, Terveen and McDonald described "social matching systems" for the scenario of incentivizing social interaction between yet unknown people [40]. As the first two steps, they highlighted the profiling of users and the computation of matches. Since the publication of the paper and with the rise of smartphones, new approaches are possible, and our concept proposes a fully automatized approach for these steps.

Guy et al. give an overview about recommending people to people in social media [18]. They distinguish between different types of edges in different SNSs, for example, unidirectional edges like following on Twitter or bidirectional like confirmed connections on Facebook. Furthermore, they distinguish between different relationship types, e.g., recommending *familiar*, *interesting*, or *similar* people. Describing SimCon in this taxonomy, the type of edges would be bidirectional, as in USN scenarios, people physically meet, and (offline) social interaction is incentivized. With SimCon, we are looking to recommend similar people. Guy et al. consider ad hoc networks as special cases and give an overview of some related works, some of which is cited in the following as well.

There are several papers by Mayer et al. published between 2010 and 2016 that deal with a similar topic [25, 23, 26, 27, 24, 28]. Based on a web-based survey, in [25], the authors investigate attributes that are important in scenarios of meeting strangers. Their main finding is that rare attributes that are shared are of significant importance, e.g., interest in something that not many people are interested in or attending the same college at the other side of the world. Attending some college could be tracked via the location history of the user, while the level of uncommonness of interests would likely need some global knowledge of the distribution of interests.

In further papers, Mayer et al. refined their concept by including ideas about using smartphone data to recognize opportune situations in which to suggest strangers to meet [26]. They distinguish between three different types of context: personal, social, and relational [27]. They developed a prototype with social profiles that included "skill level" and "learning/teaching needs" for a university setting [24]. The evaluation shows that "personal context (mood and busyness) and sociability of others [are] the strongest predictors of contextual match interest" [28]. The difference to SimCon is that for Mayer et al., the focus was the conceptualization of the factors that play a role when recommending contacts. We focus on the opportunities that homophilic recommendations can give by incorporating context data from a multitude of sources. SimCon could be extended by the insights gained from Mayer et al. by incorporating their findings regarding opportune situations to apply contact recommendations.

There are a few concrete studies following similar approaches. Qiao et al. developed a recommender system for recommending strangers in proximity based on similar check-in behaviors [34]. They suggest this approach as an extension to just using limited profile data. As an extension to this, we argue that the data trackable on the smartphone already reflects a large variety of interests and psychological aspects.

Chin et al. developed a recommender system for workplace and academic conference settings. In [10], they present a study that concludes that already knowing each other and having met before are the top two reasons to add a new contact. In additional papers, the authors report about multiple experiments in workplace and conference settings that confirm those findings [9, 8].

In more recent studies (2017 and 2018), Xia et al. and Asabere et al. used similarity in social tie connection, i.e., contact duration and contact frequency during a conference, as well as similarity in personality traits, for a contact recommender system [42, 1]. The major difference to our work is the work-related scenario, in which people will often meet the same people, and the overall number of people to meet is limited. Furthermore, in workplaces, the purpose of meeting, most likely, is focused on work relationships. SimCon, in contrast, is more generic and also applies to scenarios where strangers meet without having to have any common work background or previous interactions.

Link Prediction One of the common ways to research links in social networks is *link prediction*. The key difference to our work is that here, the social graph is used to calculate the prediction or recommendation, while SimCon is also feasible in USN scenarios where no social graph is available. Yin et al. analyzed links in social networks based on "intuition-based" aspects: homophily (shared attributes), rarity (matching uncommon attributes), social influence (more likely to link to person who shares attributes with existing friends), common friendship (mutual friends), social closeness (being close to each other in the social graph), and preferential attachment (more likely to link to a popular person) [44]. Most aspects focus on the social graph or global knowledge about attribute distribution (in the case of rarity). In the work by Mohajureen et al., the authors use the attributes of neighboring nodes in the friend recommendation process [29]. For this algorithm to work, the social graph and the features of each user have to be available.

9.5 Summary

In this chapter, we proposed a concept for contact recommendations in USN scenarios—SimCon. We showed that research results from psychology and social sciences suggest that we can calculate relevant contact recommendations based on smartphone data, without utilizing the social graph. Our approach SimCon can be applied fully automatized, from data collection, over data exchange, to similarity estimation and contact recommendation.

After collecting smartphone data, in order to estimate the similarity of two users, they can share their data in a device-to-device manner by utilizing appropriate data structures and appropriate wireless interfaces. In the next chapter, we will introduce our solution to similarity estimation for USN scenarios. Furthermore, in Chap. 11, we will give details about implementing seamless device-to-device communication on off-the-shelf smartphones.

References

1. N.Y. Asabere, A. Acakpovi, M.B. Michael, Improving socially-aware recommendation accuracy through personality. IEEE Trans. Affect. Comput. 9(3), 351–361 (2018). https://doi.org/10.1109/TAFFC.2017.2695605
2. M.D. Back, J.M. Stopfer, S. Vazire, S. Gaddis, S.C. Schmukle, B. Egloff, S.D. Gosling, Facebook profiles reflect actual personality not self-idealization. Psychol. Sci. 21(3), 372–374 (2010). https://doi.org/10.1177/0956797609360756
3. J. Bao, Y. Zheng, D. Wilkie, M. Mokbel, Recommendations in location-based social networks: a survey. Geoinformatica 19(3), 525–565 (2015). https://doi.org/10.1007/s10707-014-0220-8
4. A. Beach, M. Gartrell, S. Akkala, J. Elston, J. Kelley, K. Nishimoto, B. Ray, S. Razgulin, K. Sundaresan, B. Surendar, M. Terada, R. Han, WhozThat? Evolving an ecosystem for context-aware mobile social networks. IEEE Netw. 22(4), 50–55 (2008)
5. F. Beierle, K. Grunert, S. Göndör, V. Schlüter, Towards psychometrics-based friend recommendations in social networking services, in 2017 IEEE International Conference on AI & Mobile Services (AIMS), June (IEEE, New York, 2017), pp. 105–108. https://doi.org/10.1109/AIMS.2017.22
6. D.M. Boyd, N.B. Ellison, Social network sites: definition, history and scholarship. J. Comput.-Mediat. Commun. 13(1), 210–230 (2007). https://doi.org/10.1111/j.1083-6101.2007.00393.x
7. A.C. Champion, Z. Yang, B. Zhang, J. Dai, D. Xuan, D. Li, E-SmallTalker: a distributed mobile system for social networking in physical proximity. IEEE Trans. Parall. Distrib. Syst. 24(8), 1535–1545 (2013). https://doi.org/10.1109/TPDS.2012.251
8. A. Chin, Ephemeral social networks, in Mobile Social Networking: An Innovative Approach, ed. by A. Chin, D. Zhang. Computational Social Sciences (Springer, New York, 2014), pp. 25–64. https://doi.org/10.1007/978-1-4614-8579-7_3
9. A. Chin, B. Xu, F. Yin, X. Wang, W. Wang, X. Fan, D. Hong, Y. Wang, Using proximity and homophily to connect conference attendees in a mobile social network, in 2012 32nd International Conference on Distributed Computing Systems Workshops, June 2012, pp. 79–87. https://doi.org/10.1109/ICDCSW.2012.56
10. A. Chin, B. Xu, H. Wang, Who should I add as a "Friend"? A study of friend recommendations using proximity and homophily. Proceedings of the 4th International Workshop on Modeling Social Media, MSM '13. Association for Computing Machinery, May 2013, pp. 1–7. https://doi.org/10.1145/2463656.2463663
11. N. Confessore, D. Hakim, Data Firm Says 'Secret Sauce' Aided Trump; Many Scoff. New York Times (March 2017)
12. R. Farahbakhsh, X. Han, A. Cuevas, N. Crespi, Analysis of publicly disclosed information in Facebook profiles, in Proceedings of the 2013 IEEE/ACM International Conference on Advances in Social Networks Analysis and Mining (ASONAM), August (ACM, New York, 2013), pp. 699–705

13. S. Gambs, M.-O. Killijian, M.N.D.P. Cortez, Show me how you move and I will tell you who you are. Trans. Data Privacy **4**, 103–126 (2010). https://doi.org/10.1145/1868470.1868479
14. J. Golbeck, C. Robles, K. Turner, Predicting personality with social media, in *Proc. CHI '11 Extended Abstracts on Human Factors in Computing Systems (CHI EA'11)*, CHI EA '11, May 2011 (ACM, New York, 2011), pp. 253–262. https://doi.org/10.1145/1979742.1979614
15. S. Göndör, F. Beierle, S. Sharhan, H. Hebbo, E. Kücükbayraktar, A. Küpper, SONIC: bridging the gap between different online social network platforms, in *Proceedings of the 2015 IEEE International Conference on Smart City/SocialCom/SustainCom (SmartCity)*, December 2015 (IEEE, New York, 2015), pp. 399–406. https://doi.org/10.1109/SmartCity.2015.104
16. S.D. Gosling, S. Gaddis, S. Vazire, Personality impressions based on facebook profiles, in *Proceedings of the International AAAI Conference on Weblogs and Social Media (ICWSM)* (AAAI, Palo Alto, 2007), pp. 1–4
17. S. Greenwood, A. Perrin, M. Duggan, *Social Media Update 2016*. November 2016. http://www.pewinternet.org/2016/11/11/social-media-update-2016/
18. I. Guy, People recommendation on social media. in *Social Information Access: Systems and Technologies*, ed. by P. Brusilovsky, D. He. Lecture Notes in Computer Science (Springer International Publishing, New York, 2018), pp. 570–623. https://doi.org/10.1007/978-3-319-90092-6_15
19. J. Heidemann, M. Klier, F. Probst, Online social networks: a survey of a global phenomenon. Comput. Netw. **56**(18), 3866–3878 (2012). https://doi.org/10.1016/j.comnet.2012.08.009
20. M. Kosinski, D. Stillwell, T. Graepel, Private traits and attributes are predictable from digital records of human behavior. Proc. Natl. Acad. Sci. **110**(15), 5802–5805 (2013). https://doi.org/10.1073/pnas.1218772110
21. C.A. Lampe, N. Ellison, C. Steinfield, A familiar face(book): profile elements as signals in an online social network, in *Proceedings of the SIGCHI Conference on Human Factors in Computing Systems (CHI 2007)*, April 2007 (ACM, New York, 2007), pp. 435–444. https://doi.org/10.1145/1240624.1240695
22. H. Liu, Social network profiles as taste performances. J. Comput.-Mediat. Commun. **13**(1), 252–275 (2007). https://doi.org/10.1111/j.1083-6101.2007.00395.x
23. J. Mayer, Is there a place for serendipitous introductions? in *Proceedings of the Companion Publication of the 17th ACM Conference on Computer Supported Cooperative Work & Social Computing*. CSCW Companion '14 (ACM, New York, 2014), pp. 73–76. https://doi.org/10.1145/2556420.2556827
24. J. Mayer, Q. Jones, Encount'r: exploring passive context-awareness for opportunistic social matching. in *Proceedings of the 19th ACM Conference on Computer Supported Cooperative Work and Social Computing Companion*. CSCW '16 Companion (ACM, New York, 2016), pp. 349–352. https://doi.org/10.1145/2818052.2869080
25. J.M. Mayer, S. Motahari, R.P. Schuler, Q. Jones, Common attributes in an unusual context: predicting the desirability of a social match, in *Proceedings of the Fourth ACM Conference on Recommender Systems*. RecSys '10 (ACM, New York, 2010), pp. 337–340. https://doi.org/10.1145/1864708.1864781
26. J.M. Mayer, S.R. Hiltz, Q. Jones, in *Making Social Matching Context-Aware: Design Concepts and Open Challenges* (ACM Press, New York, 2015), pp. 545–554. https://doi.org/10.1145/2702123.2702343
27. J.M. Mayer, Q. Jones, S.R. Hiltz, Identifying opportunities for valuable encounters: toward context-aware social matching systems. ACM Trans. Inf. Syst. **34**(1), 1:1–1:32 (2015). https://doi.org/10.1145/2751557
28. J.M. Mayer, S.R. Hiltz, L. Barkhuus, K. Väänänen, Q. Jones, Supporting opportunities for context-aware social matching: an experience sampling study, in *Proceedings of the 2016 CHI Conference on Human Factors in Computing Systems*. CHI '16. Association for Computing Machinery May 2016, pp. 2430–2441. https://doi.org/10.1145/2858036.2858175

29. M. Mohajireen, C. Ellepola, M. Perera, I. Kahanda, U. Kanewala, Relational similarity model for suggesting friends in online social networks, in *Proceedings of the 2011 6th International Conference on Industrial and Information Systems (ICIIS)*, August (IEEE, New York, 2011), pp. 334–339. https://doi.org/10.1109/ICIINFS.2011.6038090
30. J. Müller, C. Anneser, M. Sandstede, L. Rieger, A. Alhomssi, F. Schwarzmeier, B. Bittner, I. Aslan, E. André, Honeypot: a socializing app to promote train commuters' wellbeing, in *Proceedings of the 17th International Conference on Mobile and Ubiquitous Multimedia. MUM 2018* (ACM, New York, 2018), pp. 103–108. https://doi.org/10.1145/3282894.3282901
31. T.H. Nguyen, D.Q. Tran, G.M. Dam, M.H. Nguyen, Estimating the similarity of social network users based on behaviors. Vietnam J. Comput. Sci. **5**(2), 165–175 (2018). https://doi.org/10.1007/s40595-018-0112-1
32. G.S. Njoo, K.-W. Hsu, W.-C. Peng, Distinguishing friends from strangers in location-based social networks using co-location, in *Pervasive and Mobile Computing*, vol. 50 (October 2018), pp. 114–123. https://doi.org/10.1016/j.pmcj.2018.09.001
33. T. Paul, A. Famulari, T. Strufe, A survey on decentralized online social networks. Comput. Netw. **75**, Part A, 437–452 (2014). https://doi.org/10.1016/j.comnet.2014.10.005
34. X. Qiao, W. Yu, J. Zhang, W. Tan, J. Su, W. Xu, J. Chen, Recommending nearby strangers instantly based on similar check-in behaviors. IEEE Trans. Autom. Sci. Eng. **12**(3), 1114–1124 (2015). https://doi.org/10.1109/TASE.2014.2369429
35. A. Richter, M. Koch, Functions of social networking services, in *Proceedings of the 8th International Conference on the Design of Cooperative Systems (COOP '08)*, May 2008, pp. 87–98
36. V.A. Rohani, O.S. Hock, On social network web sites: definition, features, architectures and analysis tools. J. Comput. Eng. **1**, 3–11 (2009)
37. P. Sapiezynski, A. Stopczynski, D.K. Wind, J. Leskovec, S. Lehmann, Offline behaviors of online friends (November 2018). arXiv:1811.03153 [cs]
38. J. Teng, B. Zhang, X. Li, X. Bai, D. Xuan, E-Shadow: lubricating social interaction using mobile phones. in *2011 31st International Conference on Distributed Computing Systems*, June 2011, pp. 909–918. https://doi.org/10.1109/ICDCS.2011.48
39. J. Teng, B. Zhang, X. Li, X. Bai, D. Xuan, E-Shadow: lubricating social interaction using mobile phones. IEEE Trans. Comput. **63**(6), 1422–1433 (2014). https://doi.org/10.1109/TC.2012.290
40. L. Terveen, D.W. McDonald, Social matching: a framework and research agenda. ACM Trans. Comput.-Hum. Interact. (TOCHI) **12**(3), 401–434 (2005). https://doi.org/10.1145/1096737.1096740
41. D. Wang, D. Pedreschi, C. Song, F. Giannotti, A.-L. Barabasi, Human mobility social ties, and link prediction, in *Proceedings of the 17th ACM SIGKDD International Conference on Knowledge Discovery and Data Mining KDD '11* (ACM, New York, 2011), p. 1100. https://doi.org/10.1145/2020408.2020581
42. F. Xia, N.Y. Asabere, H. Liu, Z. Chen, W. Wang, Socially aware conference participant recommendation with personality traits. IEEE Syst. J. **11**(4), 2255–2266 (2017). https://doi.org/10.1109/JSYST.2014.2342375
43. Z. Yang, B. Zhang, J. Dai, A. Champion, D. Xuan, D. Li, E-SmallTalker: a distributed mobile system for social networking in physical proximity, in *2010 IEEE 30th International Conference on Distributed Computing Systems (ICDCS)*, June (IEEE, New York, 2010), pp. 468–477. https://doi.org/10.1109/ICDCS.2010.56
44. Z. Yin, M. Gupta, T. Weninger, J. Han, A unified framework for link recommendation using random walks, in *Proceedings of the 2010 International Conference on Advances in Social Networks Analysis and Mining (ASONAM)* (IEEE, New York, 2010), pp. 152–159. https://doi.org/10.1109/ASONAM.2010.27

Chapter 10
Similarity Estimation

In this chapter, we address B-Task3 (*Develop a concept to assess the similarity of users in ubiquitous social networking scenarios.*) and develop metrics for the estimation of the similarity of two users based on their smartphone data. While approaches for similarity estimation exist for sets, our metrics apply to multisets and thus can estimate similarities in frequencies of events, for example, playback events of music listened to. A previous version of the results presented in this chapter has been published in [3].

10.1 Motivation

In January of 2018, the prime minister of the UK appointed one of her ministers to focus on issues related to loneliness.[1] This acknowledges the basic human need to form connections with other people. From a technological point of view, SNSs (social networking services) are typically used to reflect real-world social connections and establish new ones. Besides established global SNSs, there is a rising market for services that explicitly focus on the connections between people in proximity, i.e., neighborhood networks like Nextdoor[2] and its competitors. We note the apparent desire and need for people to form connections with those in physical proximity. In the previous chapter, we already started to show how technology can assist this need.

First, we observe that people use their smartphone to access SNSs. Online social networks have shifted to mobile social networks. Facebook, for instance, lists 1.15

---

[1] https://www.theguardian.com/society/2018/jan/16/may-appoints-minister-tackle-loneliness-issues-raised-jo-cox.

[2] http://www.nextdoor.com/.

F. Beierle, *Integrating Psychoinformatics with Ubiquitous Social Networking*,
T-Labs Series in Telecommunication Services,
https://doi.org/10.1007/978-3-030-68840-0_10

billion mobile daily active users on average for 2016.[3] Among the most frequent concerns with established, centralized SNSs are the potential misuse of data and loss of user privacy. This concern is heightened by the highly sensitive user and sensor data that modern smartphones provide (cf. first part of this thesis). In our discussions about decentralization in Sect. 8.2 (page 84), we argued how decentralized systems can help mitigate privacy concerns and lock-in effects posed by centralized systems. By utilizing short-range wireless interfaces, the need for centralized servers in social networking scenarios can be reduced. Utilizing Bluetooth, WiFi Direct, or NFC, smartphones can communicate directly with each other. Section 8.2 has shown the prevalence of device-to-device communication for USN (ubiquitous social networking) scenarios.

We continue following the use case of incentivization of social interaction between strangers and build on the results of Chap. 9. By utilizing device-to-device communication, we address the privacy concerns related to centralized SNS architectures. Following the results of the previous chapter, we are aiming at determining the similarity of smartphone data between two users.

We have to consider the applicability of the solution in a quickly changing mobile device-to-device context, as well as implications imposed by short-range wireless technologies. This implies the use of only small amounts of data and a limited number of necessary data exchanges. As the needed data is already present on the smartphone and connectivity between devices is also available, we focus on the question of how to process and compare the data under the constraints of bandwidth and computing limitations of mobile devices.

Based on the use case of two strangers meeting, we develop a method of similarity estimation for USN scenarios. Two users can quickly approximate their similarity when meeting, without exchanging clear text data nor contacting any central server. While our approach is applicable to any other data that can be represented as a multiset, in this chapter, we focus on one of the most typical features of social profiles, the musical taste of the user. Not only is listening to music one of the most typical usages for smartphones;[4] musical taste is, after gender, the most commonly disclosed profile feature in Facebook [13]. Musical taste is a common feature that users tend to identify with, which thus can serve as an appropriate feature for similarity estimation or can be used as an example in the recommendation of new contacts.

In this chapter, we present our approach that allows the estimation of the similarity of two users' musical tastes based on probabilistic data structures. While approaches for set similarity estimation exist for the Bloom filter (BF), we develop an approach for Counting Bloom filters (CBFs) and Count-Min Sketches (CMSs), suitable for multisets. We discuss our approach based on experiments done with synthetic data and real user music listening history data. We conclude with a

---

[3] https://investor.fb.com/investor-news/press-release-details/2017/Facebook-Reports-Fourth-Quarter-and-Full-Year-2016-Results/default.aspx.

[4] http://www.pewinternet.org/2015/04/01/us-smartphone-use-in-2015/.

concrete approach that is applicable for multiset similarity estimations in device-to-device scenarios.

10.2 Existing Similarity Assessment Approaches

Most projects reviewed in the related work section (Sect. 8.2) let the user manually evaluate his/her interest in another user's profile without automatic similarity estimation or calculation (e.g., [11, 2, 22, 25, 26, 7, 21]). However, there are also some automated approaches that we will review here.

In *SANE*, the devices of users with similar interests are used for forwarding messages. User similarity is calculated automatically. The whole network has a fixed space of interests, and each user has an interest profile with weights for each possible item in the interest space [19, 20]. Luo et al. [18] follows the same approach for interest profiles. In [17], a similar idea is followed, just without weighting of interests—here, interest in each item is stored in a binary fashion. The difference in our approach is that we work with an interest/preference space that is vast and cannot be predetermined: the space of all possible music tracks is constantly expanding.

E-SmallTalker employs so-called *iterative Bloom filters* for finding the intersection of two sets of topics of interest [28, 6]. In principle, the interest space here is that of any possible string. There is, however, no possibility to assign weights to the interests or order them by importance. In contrast, in our approach, we can express the frequency with which the user listened to each track, indicating how much the user likes it.

Papers that focus more on the algorithmic side of this topic often deal with the research areas of *private set intersection* or *secure multi-party computation*. Here, the application scenarios usually require a much higher level of privacy than estimating similarity in a USN scenario. Especially, the factor of having to be in physical proximity reduces potential attack vectors. Often, multiple data exchanges are necessary for handshakes, key exchanges, etc. [9]. Furthermore, some approaches rely on third parties to perform homomorphic encryption [16] or need other peers to perform computations [27, 14]. While these third parties usually do not learn anything about the two users, they are not needed in our approach, with which it is possible to estimate the similarity of two users with one single device-to-device data exchange.

10.3 Background

Multisets A *multiset* is a generalization of a set, allowing for multiple instances of each of its elements. A series of events for which the frequency is important—but the order is not—can be described as a multiset. Take, for example, the visited locations of a user. Every location or area can be described by a unique string.

Each visit adds one element of the corresponding string to the multiset. Users with similar movement patterns will generate similar multisets. In this chapter, we focus on the musical taste of users. Without the need to have the user explicitly enter this information, we can just collect data about the songs a user listened to with applications like TYDR. Storing a unique string representation for each song for each time it is played yields a multiset that represents the musical taste of the user. In order to reduce the amount of data that needs to be exchanged, we want to avoid sending clear text music playlists between clients.

Bloom Filter (BF) Probabilistic data structures are able to represent large amounts of data space efficiently. Querying the data yields results with a certain probability. The trade-off is between used memory and precision. A BF yields probabilistic set membership [5]. It consists of a bit vector with n bits, initialized with zeros, and k pairwise independent hash functions, each of which yields one position in the bit vector when hashing one element. When adding an element to the set, all hash functions are applied, yielding k positions in the bit array, which then are switched to 1. We visualize a BF in Fig. 10.1. When querying for set membership, the BF can answer whether the element is definitely not in the set—when the query element hashes to at least one 0 in the bit vector—or is likely in the set, when all hash positions in the bit vector are 1. In the latter case the queried element is either in the set or there is a collision with another element (false positive).

Counting Bloom Filter (CBF) An extension to the BF to adapt it to work with multisets is to make each field in the bit vector not binary but a counter [12]. An example is shown in Fig. 10.2. The resulting CBF can yield a close estimation of the cardinality of the queried element in a multiset. Analogously to the false positives due to collisions in the original BF, the estimated cardinality from the CBF represents an upper bound of an element's cardinality in the original multiset. In order to achieve a closer approximation of the cardinality of an element, utilizing multiple hash functions can be useful. A single hash function can have collisions, and there can be collisions between hash functions.

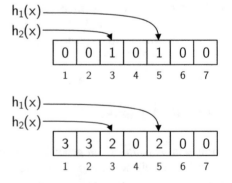

Fig. 10.1 Visualization of a Bloom filter with a length of $n = 7$ and two hash functions ($k = 2$)

Fig. 10.2 Visualization of a Counting Bloom filter with a length of $n = 7$ and two hash functions ($k = 2$)

Fig. 10.3 Visualization of a
Count-Min sketch with a
width of $w = 7$ and depth of
$d = 2$

$h_1()$	1	0	3	0	4	0	0
$h_2()$	0	4	0	0	1	0	3
	1	2	3	4	5	6	7

In this chapter, we use *collision* instead of *hash collision*, because the collisions that occur do not have to be hash collisions: as each element has to be mapped to the length of the (C)BF and not the entire namespace of the hash function, there may be collisions even if there is no hash collision. For example, two different hashes from hash function h_1 could be mapped to the same position of the (C)BF, resulting in a collision but strictly speaking without having a *hash* collision. Utilizing multiple hash functions can help reduce the impact of collisions: when querying a CBF for the cardinality of an item e, it is hashed with each hash function, and the lowest counter is returned. Through the described collisions, the yielded number is equal or higher than the true amount.

Count-Min Sketch (CMS)   Another probabilistic data structure that works with multisets is the Count-Min Sketch [8]. It is often used to provide information about the frequency of events in streams of data. A CMS consists of w columns and d rows (cf. Fig. 10.3). Each field is initialized with 0. Each row is associated with one hash function. Adding an element increments the counter at the positions indicated by the hash functions.

It is worth stating the relationship between CBF and CMS: adding the rows of a CMS yields a CBF. A CBF with length n and one single ($k = 1$) hash function h is equal to a CMS with width $n = w$ and depth 1, given that the same hash function h is used in the CMS:

$$CBF_{n,k} = CMS_{w,d} \quad | \quad n = w \quad \wedge \quad k = d = 1 \tag{10.1}$$

Comparison of Two BFs   There are some papers dealing with the comparison of two sets utilizing BFs [15, 23, 10, 1]. In [15], the authors compare two BFs with a bit-wise *AND*. A large number of 1s in the result indicates similarity. The authors of [23] calculate string similarity utilizing BFs by creating n-grams, adding those into BFs and using the Dice coefficient (see Eq. (10.3)). The idea is that identical n-grams will hash to identical positions in the bit array as long as the same length and same hash functions are used. In [10], the authors utilize BFs to estimate path similarity for paths in computer networks. They define a *Bloom Distance*, which is the logical *AND* of both BFs, followed by counting the number of 1s in the results and dividing by the length of a BF. The authors of [1] use cosine similarity on two BFs to determine similarity. In [28], the authors iteratively compare two BFs. They start with a small BF with a high false-positive rate and increase its size in a second round of comparisons if the similarity value of the first round was above a predefined threshold.

10.4   Metrics for Comparing CBFs and CMSs

Because in our scenario, we are dealing with multisets, utilizing a BF and thus leaving out the cardinality of each element in the multiset would not reflect the musical taste of the user anymore. It is the cardinality of the element ("play count") that indicates the number of times a specific song was played back.

In order to compare multisets, one can apply similar metrics like suggested for the BF, i.e., cosine similarity and Dice coefficient. Both cosine similarity (in case of positive values) and Dice coefficient yield a value between 0 and 1, where 0 indicates no similarity and 1 indicates sameness. The cosine similarity for two vectors v_1 and v_2 is defined as:

$$cosSim(v_1, v_2) = \frac{v_1 \cdot v_2}{||v_1|| \cdot ||v_2||} \tag{10.2}$$

The numerator indicates the dot product of v_1 and v_2. The denominator indicates the multiplication of the lengths of the two vectors: $||x|| = \sqrt{x_1^2 + \ldots + x_n^2}$. Given a multiset X, we can interpret the cardinalities of the elements in the multiset as elements of a vector. Thus, for two multisets X, Y, we will just write $cosSim(X, Y)$, assuming an appropriate vector representation of the cardinalities of the elements in X and Y.

The *Dice coefficient* of two sets A and B is given by:

$$Dice(A, B) = \frac{2 \cdot |A \cap B|}{|A| + |B|} \tag{10.3}$$

In order to obtain the Dice coefficient for two multisets X and Y, we can also use Eq. (10.3) by employing the cardinality and the intersection of multisets. The cardinality gives the sum of all occurrences of all elements in the multiset. The intersection of two multisets X and Y can be determined by the minimum function applied for each element i: if $i \in^n X$ (denoting that X has exactly n instances of i) and $i \in^m Y$, then the following holds for each element:

$$i \in^{min(n,m)} (X \cap Y) \tag{10.4}$$

To the best of our knowledge, there is no research specifically about the comparison of two CBFs or two CMSs. A prerequisite for the pairwise comparison is that the two data structures to be compared are of the same length and use the same hash functions; i.e., the same elements will always hash to the same positions in the data structure. Based on these prerequisites, we will now transform the idea of both cosine similarity and Dice coefficient to CBFs as well as to CMSs, yielding metrices for these data structures.

10.4.1   Cosine Similarities for CBFs and CMSs

As the data structure of a CBF is a vector, we can immediately use Eq. (10.2) for defining the cosine similarity $cosSim(P, Q)$ of two CBFs P, Q.

For the cosine similarity of CMSs, we view each CMS as a collection of d vectors (one vector each row) (cf. Fig. 10.3) and propose the following.[5]

Definition 10.1 (CMS-cosSim) Let R and S be two CMSs with the same dimensions $d \times w$ and utilizing the same hash functions. The *CMS cosine similarity* of R and S is given by:

$$CMS\text{-}cosSim(R, S) = \frac{1}{d} \cdot \sum_{i=1}^{d} cosSim(\vec{r}_i, \vec{s}_i) \qquad (10.5)$$

where $\vec{r}_i = (r_{i1}, \ldots, r_{iw})$ and $\vec{s}_i = (s_{i1}, \ldots, s_{iw})$.

Thus, the CMS cosine similarity of two CMSs averages the cosine similarity of all corresponding rows.

10.4.2   Dice Coefficients for CBFs and CMSs

The bitwise operations suggested for the comparisons of two BFs do not work with CBF and CMS as we have counters—instead of binary values—at each position (cf. Figs. 10.2 and 10.3). Therefore, we transfer the idea of the Dice coefficient for multisets to CBFs as follows:

Definition 10.2 (CBF-Dice Coefficient) Let P and Q be two CBFs with length n and utilizing the same hash functions. The *CBF-Dice coefficient* of P and Q is given by:

$$CBF\text{-}Dice(P, Q) = \frac{2 \cdot \sum_{i=1}^{n} min(p_i, q_i)}{\sum_{i=1}^{n} p_i + q_i} \qquad (10.6)$$

Note that in order to approximate the numerator for the multiset Dice coefficient, the CBF-Dice coefficient in Eq. (10.6) applies the minimum function for each position, and for the denominator, the cross sum of both CBFs is used.

Extending the Dice coefficient to CMSs reuses the Dice coefficient for CBFs.

Definition 10.3 (CMS-Dice Coefficient) Let R and S be two CMSs with the same dimensions $d \times w$ and utilizing the same hash functions. The *CMS-Dice coefficient* of R and S is given by:

---

[5] If P (resp. R) is a CBF (resp. CMS), its positions will be denoted by p_i (resp. r_{ij}).

$$CMS\text{-}Dice(R, S) = \frac{1}{d} \cdot \sum_{i=1}^{d} \frac{2 \cdot \sum_{j=1}^{w} min(r_{ij}, s_{ij})}{\sum_{j=1}^{w} r_{ij} + s_{ij}} \qquad (10.7)$$

Hence, the Dice coefficient of two CMSs utilizes the average of the CBF-Dice coefficient of each pair of corresponding rows.

10.4.3   Comparing Multisets via CBFs and CMSs

Given two multisets X and Y, we can now estimate their similarity via CBFs or via CMSs. Let X^{cbf} and Y^{cbf} be CBFs (of the same length and using the same hash functions), and let X^{cms} and Y^{cms} be CMSs (of the same dimensions and using the same hash functions) for X and Y. In the following evaluation, we show how we can use

$$CBF\text{-}Dice(X^{cbf}, Y^{cbf}) \qquad (10.8)$$

and

$$CMS\text{-}Dice(X^{cms}, Y^{cms}) \qquad (10.9)$$

to approximate

$$Dice(X, Y). \qquad (10.10)$$

Likewise, we investigate the approximation of

$$cosSim(X, Y) \qquad (10.11)$$

by

$$cosSim(X^{cbf}, Y^{cbf}) \qquad (10.12)$$

and

$$CMS\text{-}cosSim(X^{cms}, Y^{cms}). \qquad (10.13)$$

Using any of the approximations given in (10.8), (10.9), (10.12), or (10.13) will typically yield a fast and space-efficient comparison of the multisets X and Y. Our evaluation on both synthetic and on real data shows that good approximations may already be achieved when using small CBFs or CMSs.

10.5 Experimental Results

For the evaluation, we use both a synthetic dataset and real user music listening histories. The synthetic dataset

$$SD = A_r, A_0, \ldots, A_{1000}$$

consists of 1002 multisets of strings. For the strings contained in each of those multisets, we use random ASCII strings that are ten characters long. As the strings are entered into the hash functions of CBF and CMS, we could have picked any other random string to achieve the same effects. Each multiset has 66.9 unique entries on average. Given a multiset of random strings A_r, the other multisets A_i are chosen such that comparing A_r to the other multisets yields Dice coefficients of 0.000 to 1.000 in increments of 0.001, i.e.:

$$Dice(A_r, A_i) = i * 0.001$$

We also use a real dataset RD in our evaluation. The key difference is that the real dataset reflects real user behavior in terms of the distribution of playback events, i.e., how many songs were listened to how often. For the real data RD, we used the taste profile subset[6] of the Million Song Dataset [4] and first created a dataset FRD (full real data). For FRD, we chose a subset of the Million Song Dataset with appropriate data for comparisons. We chose active users who each listened to at least 50 distinct songs. FRD contains 1865 distinct users, 15,556 distinct song titles, and 277,628 recorded plays. In order to be able to visualize the results, we chose a subset of these users that yield a range of different similarity values. The resulting dataset RD consists of 88 users who listened to 63.8 distinct songs on average. There are 1955 distinct songs overall in RD. For RD, comparing each user with each other user, we get 3828 overall comparisons.

Table 10.1 gives an overview over all used datasets and their properties. We start with using SD and RD and verify our results with FRD. Note that "comparison pairs" means the overall number of pairwise comparisons conducted with the dataset. For example, RD has 88 users, so there are $n * (n - 1)/2$ possible comparison pairs, comparing each user with each other user. In the following, for visualization purposes, we sort the comparison pairs by ground truth similarity score.

---

[6]https://labrosa.ee.columbia.edu/millionsong/tasteprofile.

Table 10.1 Overview of datasets used in this evaluation

	SD	RD	FRD
Users	1,002[a]	88	1,865
Comparison pairs	1,002[a]	3,828	1,738,180
Average unique inputs	67	64	66
Unique entries overall	208	1,955	15,556
Entries overall	2,984,040	14,422	277,628

[a]Note that for SD, we did not compare each user with each other user but compared one user to all other users, yielding Dice coefficients of 0.000 to 1.000 in increments of 0.001

10.5.1 Synthetic Data SD/Comparing CMSs

For evaluating CMS-Dice on SD, we start with a CMS encoding of SD with $w = 400$ columns and $d = 10$ rows. For each $A_i \in SD$, let $A_i{}^{cms}$ be the corresponding CMS for A_i. Figure 10.4 illustrates the comparison of the Dice coefficient of the multisets as ground truth—$Dice(A_r, A_i)$—with the CMS-Dice of the CMS representation: $CMS\text{-}Dice(A_r{}^{cms}, A_i{}^{cms})$. The Dice coefficient is plotted in red. The x-axis indicates the multiset pair combination that is compared—sorted by Dice coefficient. The y-axis gives the similarity score. The blue dots represent the CMS-Dice similarity score.

In Fig. 10.5, we give the same plot for cosine similarities. We compare the cosine similarity of the multisets as ground truth—$cosDis(A_r, A_i)$—with the cosine similarity of the CMS representation: $CMS\text{-}cosSim(A_r{}^{cms}, A_i{}^{cms})$. The first observation we make is that Dice and cosine measurement yield almost identical results, which can be seen by the red lines in Figs. 10.4 and 10.5 having almost identical slopes. Therefore, for the rest of this section, we focus on the Dice coefficient. Another observation we make is that the similarity estimation by CMS-Dice is always slightly higher than or equal to the Dice coefficient ground truth—so the similarity between two multisets is always correctly estimated or overestimated due to collisions, never underestimated.

Typically, when using a CMS, the user wants to perform queries and get accurate results. The values for w and d are chosen accordingly. In our case, we just want to perform a similarity estimation without querying for specific entries. We investigate what influence the number of columns w and the number of rows d have on the estimation of similarity. In order to do so, we plot the same comparisons of the synthetic data for different values of w and d. For each combination, we calculate the RMSE (root-mean-square error) of the similarity estimation by CMS-Dice from the Dice coefficient of the ground truth. The RMSE quantifies to what extent the similarity estimation differs from the ground truth similarity score. The lower the RMSE, the better the approximation of the Dice coefficient. Based on 100,100 comparisons, we calculate the RMSE for different combinations of w (x-axis) and d (y-axis) values. The result is given in Fig. 10.6. With an increasing number of columns, the RMSE decreases. The number of rows does not significantly influence

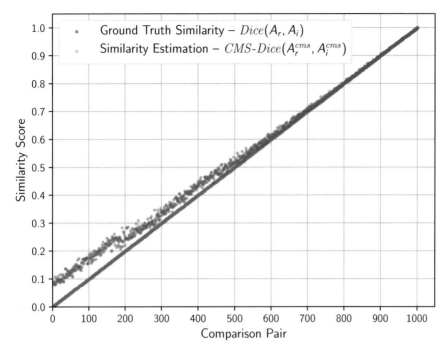

Fig. 10.4 Similarity scores for ground truth $Dice(A_r, A_i)$ in red and estimation $CMS\text{-}Dice(A_r^{cms}, A_i^{cms})$ in blue using a CMS with $w = 400$ columns and $d = 10$ rows, using synthetic dataset SD

the RMSE: increasing the number of rows does not reduce the RMSE of the similarity estimation.

10.5.2   Synthetic Data SD/Comparing CBFs

Visualizing the RMSEs for similarity estimation by CBF-Dice with different length n and number of hash functions k, we get Fig. 10.7. Increasing the length of the CBF reduces the average error of CBF-Dice while increasing the number of hash functions increases the error.

10.5.3   Real Data RD

Using the real dataset RD, our findings using synthetic data SD are confirmed. Figure 10.8 shows the RMSEs for CMS-Dice and Fig. 10.9 for CBF-Dice. We

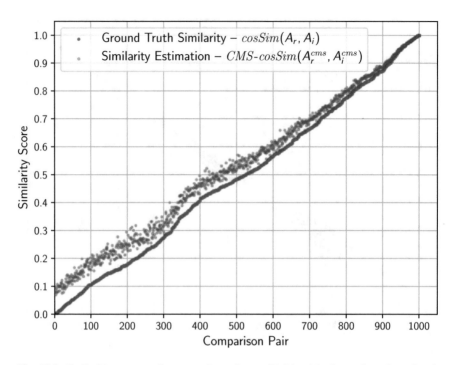

Fig. 10.5 Similarity scores for ground truth $cosDis(A_r, A_i)$ in red and estimation $CMS\text{-}cosSim(A_r{}^{cms}, A_i{}^{cms})$ in blue using a CMS with $w = 400$ columns and $d = 10$ rows, using synthetic dataset SD

achieve the highest accuracy and simultaneously the lowest memory size by using a CMS with one row, which is a CBF with one hash function (cf. Eq. (10.1)). In Fig. 10.10, we visualize the estimation error with CBF-Dice (which is equal to CMS-Dice in this case) utilizing this data structure with a length of 400. We can see the error of each similarity estimation and can see that we never underestimate the similarity. Looking at the first 3000 comparison pairs, the values for the ground truth similarity scores are very low, almost 0. The values for the similarity estimation by CBF-Dice range roughly from 0 to 0.2. For the remaining comparisons, we observe that the higher the ground truth similarity score is, the lower is the error range by the CBF-Dice estimation.

10.6 Discussion

In order to discuss the experimental results, we start by analyzing the regular BF. Consider a regular BF with one hash function, utilized for comparing sets. If two sets are equal, the BFs should be equal, and even the smallest length yields the correct

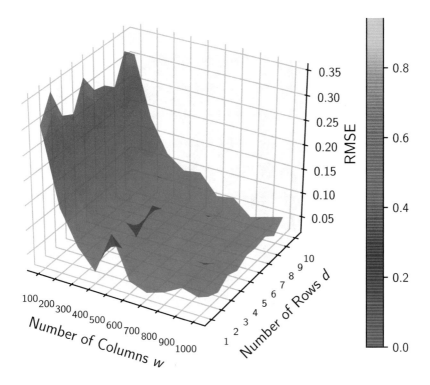

Fig. 10.6 RMSEs of similarity estimation *CMS-Dice* for CMSs of different sizes, using synthetic dataset SD and *Dice* coefficient as ground truth

estimation. Imagine comparing the BF given in Fig. 10.11a with itself. Comparing each position, the given values are always the same, yielding an estimated similarity of 1.

Regarding memory size (length of the BF), comparing two disjoint sets is the worst-case scenario: estimating the similarity of two disjoint sets by performing *AND* on the two BFs should yield 0 for every position. Imagine comparing the BFs given in Figs. 10.11a and b. The *AND* operation yields the correct estimation of 0. There is one factor that introduces a deviation from 0: the number of unique inputs for a given length of the BF. Because of the limited length of the BF, several elements are hashed to the same position in the bit vector (see description in Sect. 10.3). When performing *AND* on the two BFs in Fig. 10.11, the result is only 0 because a, b, c, and d happen to hash to the four different positions in the BF. If we add an element e to the BF on the left and it happens to hash to position 3 or 4, the comparison with the BF on the right will yield a higher value than 0 as the estimated similarity, although the exact similarity is still 0. By increasing the length of the BF, the probability for such collisions is reduced. The more unique elements are entered into the BF, the more positions are set to 1. Thus, the more unique elements are in at

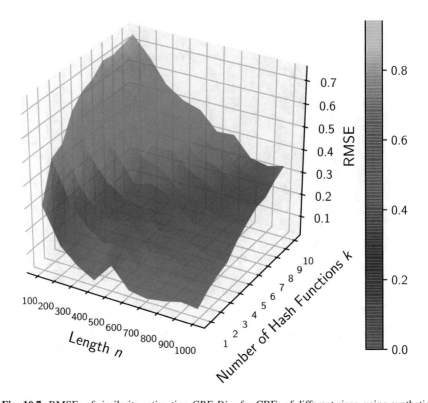

Fig. 10.7 RMSEs of similarity estimation *CBF-Dice* for CBFs of different sizes, using synthetic dataset SD and *Dice* coefficient as ground truth

least one of the sets, the longer the bit vector should be if a small error in estimating the similarity is desired.

When comparing two disjoint sets with cardinalities g and h, the BFs have to have at least a length of $g + h$ in order to theoretically be able to correctly estimate a similarity of 0 (cf. Fig. 10.11). Now imagine increasing the number of hash functions. It increases the error: more bits are set to 1, which creates a higher similarity estimation. In the following, we show that these conclusions are also true for the CMS and CBF.

As described in Sect. 10.3, when the goal is to query a CBF or CMS for the cardinality of an element of a multiset, the user profits from utilizing multiple hash functions. In our scenario of similarity estimation, we do not need to query for specific items and do not profit from multiple hash functions in the same way. As described above for the BF, the opposite is also true for the CBF: both Figs. 10.7 and 10.9 indicate the trend that the more hash functions we use, the worse the error is. This is because of the collisions: multiple elements are mapped to the same positions in the vector. The more hash functions we use, the more collisions there are. This can be compensated by increasing the length of the CBF, which would just unnecessarily increase the needed memory size.

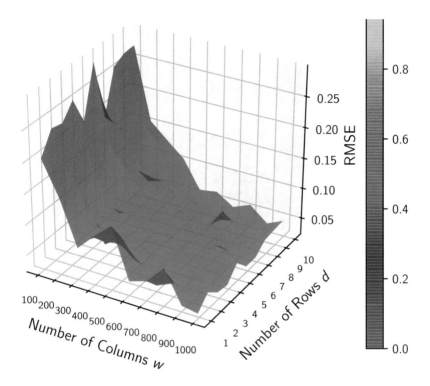

Fig. 10.8 RMSEs of similarity estimation *CMS-Dice* for CMSs of different sizes, using real dataset RD and *Dice* coefficient as ground truth.

Regarding the CMS, we increase the number of hash functions by increasing the number of rows (see Figs. 10.6 and 10.8). However, we do not see an increase in error when using more hash functions. This is because each hash function corresponds to one row. The probability of collisions is the same in each row, and the average of the row-wise similarities calculated by CMS-cosSim and CMS-Dice contains this error. Considering memory size and computation time, we should use just one single row.

The best and worst case for similarity estimation are the same as described for the BF: the higher the real similarity, the lower is the RMSE. The same elements definitely hash to the same positions. Only those elements not present in the other multiset introduce an error in the similarity estimation through collisions. Thus, the more dissimilar the multisets are, the larger the potential error. This can best be seen in Fig. 10.10. Note how the spread of blue dots (similarity estimations by CBF-Dice) spans a larger part of the *y*-axis (similarity score) for lower ground truth similarity scores (red dots). This means that for lower similarities, there are higher errors. All errors are produced by collisions and lead to an overestimation of similarity.

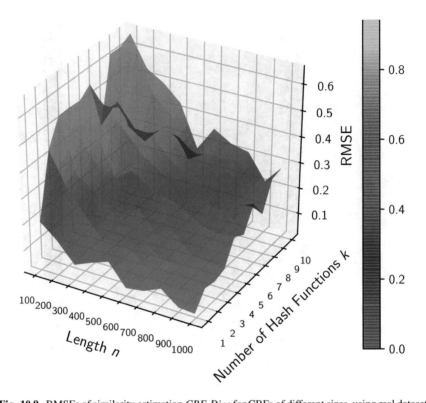

Fig. 10.9 RMSEs of similarity estimation *CBF-Dice* for CBFs of different sizes, using real dataset RD and *Dice* coefficient as ground truth

Compared to the BF comparison, for the CBF and CMS, the cardinality of each element is the additional factor to consider. For the regular BF, each collision has the same effect on the error. Using a CBF or CMS, the influence a collision has on the error of the similarity estimation is based on how the data of the multiset is distributed. If two users listen to two different songs very frequently and those two songs are mapped to the same position in the data structure, then the error in the similarity estimation can be significant. The error is less significant if the collision occurs for two different songs the two users listened to less frequently.

We conclude that a general approach for estimating the similarity of two multisets in proximity-based mobile applications can be:

- Use one-hash CBF/one-row CMS as a data structure.
- Estimate the average number of unique input elements.
- Define an appropriate threshold for the given scenario.

We showed that the one-hash CBF gives the best estimation. Our discussion showed that the average number of unique input elements is the relevant factor for how good the estimation is. Based on this number, we can pick the length of the CBF. We

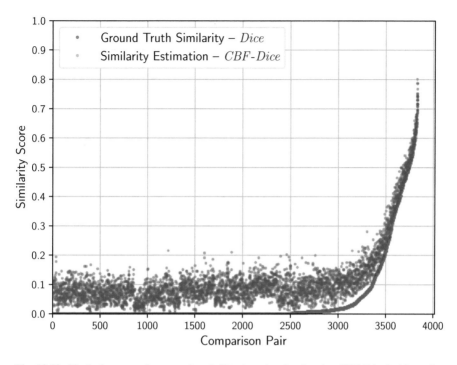

Fig. 10.10 Similarity scores for ground truth *Dice* in red and estimation *CBF-Dice* in blue using a CBF/CMS with length 400 and 1 hash function, using real dataset RD

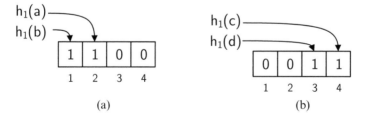

Fig. 10.11 Two Bloom filters representing the disjoint sets $\{a, b\}$ and $\{c, d\}$. (**a**) Bloom filter with elements a and b hashed to position 1 and 2. (**b**) Bloom filter with elements c and d hashed to position 3 and 4

showed that a length of twice the average unique inputs is necessary to theoretically still be able to estimate with full accuracy in the worst-case scenario of disjoint multisets. Lastly, after performing the similarity estimation, the system should define a threshold to be able to tell if the result should be regarded as significant.

Taking the music listening histories in our scenario, we have an average unique input of 63.8 elements and regard similarities above 0.6 to be relevant. Picking a size of twice the average unique input gives a length of the data structure of 128.

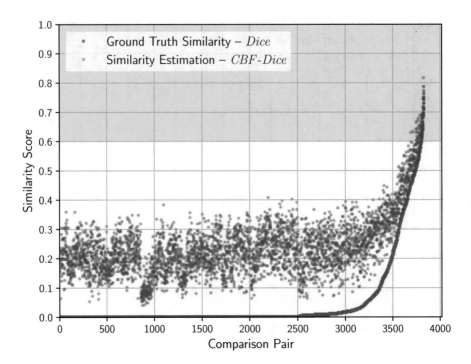

Fig. 10.12 Similarity scores for ground truth *Dice* (red) and estimation *CBF-Dice* (blue) using one-hash CBF (length 128), using real dataset RD. The green area indicates similarity scores above the desired threshold of 0.6

We plot the ground truth (red) and the CBF-Dice similarity estimation (blue) in Fig. 10.12. The green area marks similarity scores above 0.6. We observe some false positives: the blue dots in the green area that correspond to red dots below the green area. The left-most blue dot in the green area indicates the largest error. Here, we estimate a significant similarity of ≥ 0.6, while the ground truth value is about 0.5. If we need more accurate results, we choose a larger length, for example, like in Fig. 10.10. A positive side effect of the larger errors for low values is that it somewhat provides privacy through lack of accuracy: estimating a similarity of 0.4 corresponds to a ground truth value of between 0 and about 0.4—we cannot make an accurate assumption about the actual similarity.

Influence of Overall Unique Entries: Test with FRD We now evaluate to what extend the *unique entries overall* have any influence on our approach. Given that for comparison purposes, the optimal number of hash functions is 1, the length is the deciding factor, in relation to the average number of unique input elements. How well our similarity estimation works can be seen by the error it produces. This error is the difference between estimated similarity CBF-Dice and ground truth Dice similarity. In Fig. 10.12, this error is the difference between the red dots

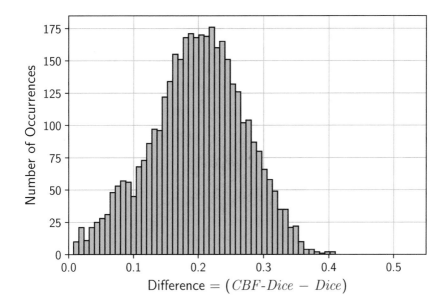

Fig. 10.13 Histogram of the differences between CBF-Dice and Dice—based on dataset RD with the properties given in Table 10.2

Table 10.2 Overview of datasets RD and FRD used for testing for influence of *unique entries overall*. The differences between CBF-Dice and Dice are plotted in Figs. 10.13 and 10.14

	RD	FRD
Users	88	1,865
Comparison pairs	3,828	1,738,180
Average unique inputs	64	66
CBF length	128	132
Unique inputs range	50–125	50–343
Unique entries overall	1,955	15,556
Entries overall	14,422	277,628

(ground truth) and the blue dots (estimated similarity). We plot a histogram of these differences in Fig. 10.13.

Looking at Table 10.2, we see that dataset FRD contains about eight times as many *unique overall entries* (15,556 vs. 1955) compared to RD. We plot the histogram of differences between CBF-Dice and Dice similarity for FRD in Fig. 10.14. The distribution of values looks almost identical compared to dataset RD. Most differences are centered around 0.2. This is evidence that the *unique entries overall*, in this case that means the unique songs in the whole dataset that is being analyzed, do not influence the performance of our approach. This, in turn, means that our approach scales and that the *average number of unique inputs* is the deciding factor when applying our similarity estimation approach. In future work, the scaling of our approach can be tested on further and also larger datasets or analyzed mathematically.

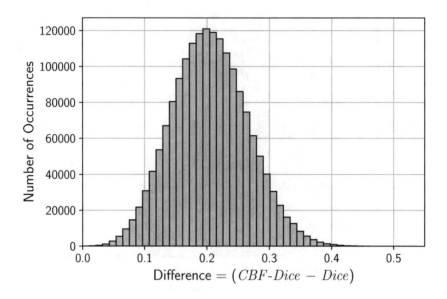

Fig. 10.14 Histogram of the differences between CBF-Dice and Dice—based on dataset FRD with the properties given in Table 10.2

The histograms in Figs. 10.13 and 10.14 show the difference between CBF-Dice similarity estimation and Dice ground truth similarity for real music listening data. In an actual deployment, the number of average unique input elements has to be estimated. What the difference between similarity estimation and real similarity looks like also depends on the distribution of the data. For example, our approach will likely perform better if all users listened to 64 unique songs than if half of the users listened to 0 unique songs and the other half listened to 128 unique songs, because the latter half of users will have more collisions when the similarity estimation is performed. Note that in FRD, the users listened to between 50 and 343 unique songs (cf. *unique input range* in Table 10.2). When implementing our approach, the system designer has to estimate the average number of unique input elements and estimate what the distribution looks like. A longer CBF can increase the accuracy of the similarity estimation if there are users expected to have significantly more unique input elements compared to the average.

The actual space needed for a CBF depends on the implementation. It can be reduced by utilizing compression. Compared to raw data transfer of song information and play counts, in our approach, we only need to transfer one CBF, i.e., a list of integer values. When implementing our approach for a USN scenario, software developers can limit the number of songs to consider. In order for the unique songs a user listened to not increase unlimited over time, only the last 64 songs could be considered, or the considered timeframe could be limited to the last few weeks or months.

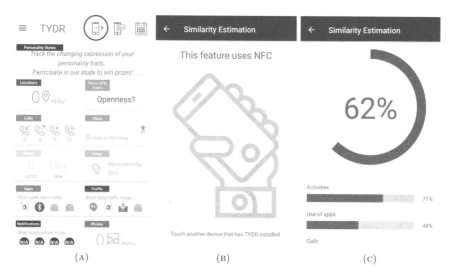

Fig. 10.15 TYDR similarity estimation extension. (**a**) TYDR extension main screen. (**b**) Waiting to connect with another user. (**c**) Result screen of the prototype

10.7 Prototypical Implementation for USN Scenarios as a TYDR Extension

In the course of his bachelor's thesis, Robert Staake prototypically implemented our approach for similarity estimation in TYDR [24]. The extended version of TYDR uses a combination of NFC and Bluetooth for encrypted data exchange between two phones and compares all the collected data points of TYDR with the other user. Figure 10.15 shows screenshots of the interaction. Imagine two users Alice and Bob meeting. They choose to exchange their TYDR data. Alice taps the button of the TYDR extension (see the red circle in Fig. 10.15a). This opens the screen shown in Fig. 10.15b. Bob only needs to bring his NFC-enabled phone in range of Alice's phone to exchange their data. Via NFC, a randomly created symmetric key is exchanged and used to establish an encrypted Bluetooth connection between the phones. Then the CBFs are exchanged and compared, and the results displayed (cf. Fig. 10.15c).

10.8 Summary

In this chapter, we approached the problem of multiset similarity estimation in USN scenarios. We developed the comparison metrics CBF-Dice and CMS-Dice for the similarity estimation of two CBFs and two CMSs. Applying these metrics, we can approximate the Dice coefficient when comparing two multisets. We evaluated our approach with both synthetic data and real music listening history data.

Our results show that the more hash functions we utilize in a data structure, the higher is the error in the estimation. The larger the data structure is, the smaller is the error. We achieve the lowest error when utilizing a one-hash CBF/one-row CMS. Here, we minimize the number of collisions by using only one hash function. The collisions are the source of the error in the similarity estimation. In general, the higher the real similarity, the better is the estimation.

The data structure one-hash CBF is appropriate for the given scenario of similarity estimation between two users, requiring only a single data exchange between two smartphones. We described the general approach for assessing the appropriate size of the data structure by estimating the average number of unique input elements as well as defining a threshold for the similarity score. Using the real user music listening histories with a mean of about 64 unique entries, we showed that a one-hash CBF with length 128 (twice the average unique inputs) suffices to accurately estimate the similarity between two multisets.

While we presented the scenario of two strangers meeting and quickly determining the similarity of their musical tastes, our approach can be applied in a variety of other scenarios. In general, any two systems that log any events can be compared with our approach. Utilizing CBF-Dice, one can perform fast and space-efficient similarity estimations of the two systems in terms of the frequencies of the logged events. We implemented the CBF-based similarity estimation as a proof of concept as an extension of TYDR.

For future work in the USN scenario, potential attack scenarios like malicious users should be addressed. Additionally, compression and encryption can help increase space efficiency and increase privacy.

The similarity of users not only is relevant for recommending new contacts but is also at the core of the idea of collaborative filtering: similar users will like similar items. In the next chapter, we will follow this idea and propose a mobile platform for decentralized recommender systems that integrates the results from this chapter.

References

1. M. Alaggan, S. Gambs, A.-M. Kermarrec, BLIP: non-interactive differentially-private similarity computation on Bloom filters, in *Stabilization, Safety, and Security of Distributed Systems*, ed. by A.W. Richa, C. Scheideler. Lecture Notes in Computer Science, vol. 7596 (Springer, Berlin, Heidelberg, 2012), pp. 202–216
2. A. Beach, M. Gartrell, S. Akkala, J. Elston, J. Kelley, K. Nishimoto, B. Ray, S. Razgulin, K. Sundaresan, B. Surendar, M. Terada, R. Han, WhozThat? Evolving an ecosystem for context-aware mobile social networks. IEEE Network **22**(4), 50–55 (2008)
3. F. Beierle, Do you like What I like? Similarity estimation in proximity-based mobile social networks, in *2018 17th IEEE International Conference On Trust, Security And Privacy In Computing And Communications (TrustCom)*, August (IEEE, New York, 2018), pp. 1040–1047. https://doi.org/10.1109/TrustCom/BigDataSE.2018.00146

4. T. Bertin-Mahieux, D.P. Ellis, B. Whitman, P. Lamere, The million song dataset, in *Proceedings of the 12th International Society for Music Information Retrieval Conference (ISMIR)* (2011)
5. B.H. Bloom, Space/time trade-offs in hash coding with allowable errors. Commun. ACM **13**(7), 422–426 (1970). https://doi.org/10.1145/362686.362692
6. A.C. Champion, Z. Yang, B. Zhang, J. Dai, D. Xuan, D. Li, E-SmallTalker: a distributed mobile system for social networking in physical proximity. IEEE Trans. Parall. Distrib. Syst. **24**(8), 1535–1545 (2013). https://doi.org/10.1109/TPDS.2012.251
7. S. Cho, C. Julien, Chitchat: navigating tradeoffs in device-to-device context sharing, in *2016 IEEE International Conference on Pervasive Computing and Communications (PerCom)*, March (IEEE, New York, 2016), pp. 1–10. https://doi.org/10.1109/PERCOM.2016.7456512
8. G. Cormode, S. Muthukrishnan, An improved data stream summary: the count-min sketch and its applications, in *LATIN 2004: Theoretical Informatics* (Springer, New York, 2004), pp. 29–38
9. C. Dong, L. Chen, Z. Wen, When private set intersection meets big data: an efficient and scalable protocol, in *Proceedings of the 2013 ACM SIGSAC Conference on Computer & Communications Security*. CCS '13 (ACM, New York, 2013), pp. 789–800. https://doi.org/10.1145/2508859.2516701
10. B. Donnet, B. Gueye, M.A. Kaafar, Path similarity evaluation using bloom filters. Comput. Netw. **56**(2), 858–869 (2012)
11. N. Eagle, A. Pentland, Social serendipity: mobilizing social software. Pervas. Comput. IEEE **4**(2), 28–34 (2005)
12. L. Fan, P. Cao, J. Almeida, A.Z. Broder, Summary cache: a scalable wide-area web cache sharing protocol. IEEE/ACM Trans. Netw. **8**(3), 281–293 (2000). https://doi.org/10.1109/90.851975
13. R. Farahbakhsh, X. Han, A. Cuevas, N. Crespi, Analysis of publicly disclosed information in Facebook profiles, in *Proceedings of the 2013 IEEE/ACM International Conference on Advances in Social Networks Analysis and Mining (ASONAM)*, August (ACM, New York, 2013), pp. 699–705
14. C.-Z. Gao, Q. Cheng, X. Li, S.-B. Xia, Cloud-assisted privacy-preserving profile-matching scheme under multiple keys in mobile social network. Clust. Comput. **22**(1), 1655–1663 (2019). https://doi.org/10.1007/s10586-017-1649-y
15. N. Jain, M. Dahlin, R. Tewari, Using bloom filters to refine web search results, in *WebDB* (2005), pp. 25–30
16. F. Kerschbaum, Outsourced private set intersection using homomorphic encryption, in *Proceedings of the 7th ACM Symposium on Information, Computer and Communications Security*. ASIACCS '12 (ACM, New York, 2012), pp. 85–86. https://doi.org/10.1145/2414456.2414506
17. T.H. Luan, R. Lu, X. Shen, F. Bai, Social on the road: enabling secure and efficient social networking on highways. IEEE Wirel. Commun. **22**(1), 44–51 (2015). https://doi.org/10.1109/MWC.2015.7054718
18. E. Luo, Q. Liu, J.H. Abawajy, G. Wang, Privacy-preserving multi-hop profile-matching protocol for proximity mobile social networks. Fut. Gener. Comput. Syst. **68**, 222–233 (2017). https://doi.org/10.1016/j.future.2016.09.013
19. A. Mei, G. Morabito, P. Santi, J. Stefa, Social-aware stateless forwarding in pocket switched networks, in *2011 Proceedings IEEE INFOCOM*, April 2011, pp. 251–255. https://doi.org/10.1109/INFCOM.2011.5935076
20. A. Mei, G. Morabito, P. Santi, J. Stefa, Social-aware stateless routing in pocket switched networks. IEEE Trans. Parall. Distrib. Syst. **26**(1), 252–261 (2015). https://doi.org/10.1109/TPDS.2014.2307857
21. J. Müller, C. Anneser, M. Sandstede, L. Rieger, A. Alhomssi, F. Schwarzmeier, B. Bittner, I. Aslan, E. André, Honeypot: a socializing app to promote train commuters' wellbeing. *Proceedings of the 17th International Conference on Mobile and Ubiquitous Multimedia*. MUM 2018 (ACM, New York, 2018), pp. 103–108. https://doi.org/10.1145/3282894.3282901

22. A.-K. Pietiläinen, E. Oliver, J. LeBrun, G. Varghese, C. Diot, MobiClique: middleware for mobile social networking, in *Proceedings of the 2nd ACM Workshop on Online Social Networks*. WOSN '09 (ACM, New York, 2009), pp. 49–54. https://doi.org/10.1145/1592665.1592678
23. R. Schnell, T. Bachteler, J. Reiher, Privacy-preserving record linkage using bloom filters. BMC Med. Inf. Dec. Making **9**(1), 41 (2009). https://doi.org/10.1186/1472-6947-9-41
24. R. Staake, Social Profile Exchanges in Proximity-based Mobile Social Networks. Bachelor's Thesis. Technische Universität Berlin, 2018
25. J. Teng, B. Zhang, X. Li, X. Bai, D. Xuan, E-Shadow: lubricating social interaction using mobile phones, in *2011 31st International Conference on Distributed Computing Systems*, June 2011, pp. 909–918. https://doi.org/10.1109/ICDCS.2011.48
26. J. Teng, B. Zhang, X. Li, X. Bai, D. Xuan, E-Shadow: lubricating social interaction using mobile phones. IEEE Trans. Comput. **63**(6), 1422–1433 (2014). https://doi.org/10.1109/TC.2012.290
27. J. Tillmanns, Privately computing set-union and set-intersection cardinality via bloom filters, in *Information Security and Privacy: 20th Australasian Conference, ACISP 2015, Brisbane, QLD, June 29–July 1, 2015, Proceedings*, vol. 9144 (Springer, New York, 2015), pp. 413–430
28. Z. Yang, B. Zhang, J. Dai, A. Champion, D. Xuan, D. Li, E-SmallTalker: a distributed mobile system for social networking in physical proximity, in *2010 IEEE 30th International Conference on Distributed Computing Systems (ICDCS)*, June (IEEE, New York, 2010), pp. 468–477. https://doi.org/10.1109/ICDCS.2010.56

Chapter 11
MobRec: Mobile Platform for Decentralized Recommender Systems

In this chapter, we address B-Task4 (*Develop example ubiquitous social networking applications taking into account the results of the previous tasks.*). As an example USN (ubiquitous social networking) application, we design, implement, and evaluate a mobile platform that enables decentralized recommender systems, MobRec. Our system implements unobtrusive device-to-device communication on off-the-shelf smartphones (Android and iOS) and does not exhibit the lock-in effects present in centralized service providers for recommender systems. Some of the results presented in this chapter have been published in [6, 14, 5].

11.1 Motivation

Recommender systems are ubiquitously available. They recommend items from different domains, for example, media to consume (e.g., Spotify, Netflix) or points of interests (POIs) to visit (e.g., Yelp, Google Maps). However, existing recommender systems have several drawbacks. Existing providers typically operate in a centralized manner: the service provider holds all the data and recommends items based on algorithms that are not visible to the user. This can introduce certain limitations and biases. Limitations often are that only items that are available with the service provider will be recommended, e.g., Netflix will only recommend items available in their catalog. Possible biases could be that the recommender algorithms favor items that create more profit for the service provider. Typically, the mentioned service providers are interested in retaining their user base and create lock-in effects. For example, movies bought on iTunes cannot be transferred to another service provider, effectively locking in the user and his/her collection.

© The Author(s), under exclusive license to Springer Nature Switzerland AG 2021
F. Beierle, *Integrating Psychoinformatics with Ubiquitous Social Networking*,
T-Labs Series in Telecommunication Services,
https://doi.org/10.1007/978-3-030-68840-0_11

The infrastructure that could be a solution for the limitations of centralized recommender systems is already in the palms of its users. The smartphone can store lots of information about its user and his/her interests, e.g., regarding preferred restaurants, music, or movies. Equipped with capabilities for device-to-device communication, users can exchange data with each other. When considering recommender systems based on content-based filtering or collaborative filtering, data about similar items and similar users are needed. Data about the properties of items can be retrieved through public APIs (e.g., Google Places, Spotify, The Movie DB (TMDB)). Finding similar users might be simple with smartphones: spending time at the same location might imply similarity—at least to a certain degree. Additionally, from our previous research, other methods of determining similarity between users based on smartphone data are available (cf. Chap. 10 and [13]). Thus, exchanging data between smartphones in proximity in a device-to-device fashion allows to create local databases that allow to filter for similar users. This data can be used for on-device recommender systems that are not limited to a single external service provider.

Combining and expanding approaches from device-to-device computing (e.g., [1, 43]) and decentralized recommender systems (e.g., [44, 3, 4]), in this chapter, we propose a modular architecture for recommender systems for virtually any domain, building on the existing infrastructure of smartphones. Our architecture consists of collaborative data collection paired with data exchange via device-to-device communication and local recommender systems running on each device, supported by third-party service providers where appropriate. There are some challenges to overcome when developing such a platform. Some data is already readily available on smartphones, for example, the most frequently visited locations. Other user preferences/ratings that cannot be assessed automatically might have to be entered manually or retrieved from external service providers, e.g., music listened to or favorite movies. Some short-distance wireless technologies, like NFC, Bluetooth, or WiFi Direct, are available on most modern smartphones, and software libraries exist for device-to-device communication. However, from an application layer perspective, utilizing such libraries for seamless data exchange between smartphones remains a challenge for multiplatform apps.

In this chapter, we propose a general modular architecture for a service-provider-independent mobile platform for recommender systems. We refer to the platform as well as our prototypical implementation as MobRec. The remainder of this chapter is structured as follows. In Sect. 11.2, we describe the requirements of MobRec. In Sect. 11.3, we give an overview of related research areas. Section 11.4 gives a detailed overview about the state of the art in device-to-device communications research and technology. A prototypical implementation of a minimum viable product (MVP) showcases the feasibility of our complete architecture. Details about the design and implementation of MobRec are given in Sects. 11.5 and 11.6. In Sect. 11.7, we evaluate MobRec, focusing on device-to-device capabilities, before concluding in Sect. 11.8.

11.2 Requirements

The general concept is that smartphone users exchange data relevant for recommender systems when they pass each other. Such a system especially works in urban areas. Today, already more than half of the world's population lives in urban areas, and the trend is that this percentage is increasing.[1] There are several things to consider when designing MobRec. In the following, we describe the technical (functional) requirements of our systems.

In order for MobRec to work, it should be deployable on off-the-shelf smartphones. Android (74.3%) and iOS (24.76%) have a combined market share of almost 100% of the global mobile operating system market.[2] The first requirement (R1) is to build a multiplatform app for both Android and iOS.

At the core of MobRec is the idea of an app running in the background, exchanging data with other users. We derive the next two requirements based on this. First, the exchange of data when two smartphones are in proximity must not require explicit user interaction in order to establish a connection (R2). If that were the case, background data exchange would no longer be possible and would require too much user effort. The next aspect is related but can be defined distinctly: the transfer of data has to be done in the background (R3). That is, the system has to be capable of transferring data from and to nearby devices while our app is running in the background. R2 and R3 can be described as the *broadcasting* of data, i.e., having a $1 : n$ relationship between sender and receiver without explicit connection establishment. R2 describes the broadcasting requirement from a user-interaction perspective, whereas R3 describes it from a technical perspective.

Considering that MobRec is a mobile platform for recommender systems, the domain of items that should be possible to be recommended is not restricted. As Sect. 11.4 will show, most device-to-device approaches that support broadcasting of messages only allow for very small payloads. Supporting a variety of different domains, in order for it to be feasible to recommend items from each supported domain, each device potentially needs to broadcast larger amounts of data (R4). As we will design a workaround for R4 in Sect. 11.5.2, we do not quantify the exact size requirements here.

To summarize, the four requirements for our system are:

R1   multiplatform (Android and iOS)
R2   no explicit connection establishment (from a user perspective)
R3   transferring data while running in the background
R4   transferring larger amounts of data.

Especially, the aspect of utilizing a background service that uses a wireless interface to broadcast and scan for messages makes battery drain a concern

---

[1] https://www.un.org/development/desa/en/news/population/2018-revision-of-world-urbanization-prospects.html.

[2] https://gs.statcounter.com/os-market-share/mobile/worldwide, accessed 2020-02-28.

that comes to mind. We focus on R1–R4 in this work. Once we designed and implemented an app with technologies that support these requirements, we propose optimizations for battery consumption in Sect. 11.7.4.

11.3   Related Research Areas

There are several areas of research that we consider related to this work. First, for research relating to the collection of data (see Part I. Second, in Sect. 8.2), we presented related work from the field of USN. The focus there is often the incentivization of social interaction, i.e., the recommendation of proximate users instead of movies, music, or POIs. Third, regarding device-to-device communication, we go into detail in the next section. Fourth, we see the concept of using decentralized systems to mitigate lock-in effects (cf. our discussions about decentralization in Sect. 8.2 (page 84)). The fifth area of related work is decentralized recommender systems.

Decentralized recommender systems traditionally use peer-to-peer networks [44, 3, 28]. Gossip protocols leverage small-world properties of overlay topologies in file sharing networks and allow to find and gather peers with similar preferences quickly and reliably. Once established, communities with similar interests exchange item ratings among each other. In contrast to such peer-to-peer scenarios, our ad hoc fashion of broadcasting and data exchange between smartphones is a network that is essentially fully disconnected. Especially, [44] and [4] follow similar approaches compared to our proposed system. In [44], decentrally stored data from the web is used for a recommender system running on the user's personal computer. Since that paper's publication (2005), the development of mobile devices enables mobile and ubiquitous scenarios depicted in this thesis. In [4], the authors propose that smartphones exchange data in a device-to-device fashion and calculate their own recommendation via collaborative filtering. The focus of that paper is on the recommender algorithm that is evaluated with a music dataset. For device-to-device communication, WiFi Direct is proposed.

Other related fields are those of ubiquitous recommender systems and context-aware recommender systems (CARS). They consider items in proximity or consider the user's current context while recommending items, respectively [23]. In contrast to these approaches, our proposed system is universal in the sense that any type of item can be recommended, independent of the item's physical proximity or the user's context.

11.4   State of the Art in Device-to-Device Communication

In this section, we give an extensive overview of related work in the field of device-to-device communications, structured by technology used. This includes

Bluetooth, WiFi, and different frameworks. The technical details given in this section might help researchers and developers in choosing appropriate device-to-device technologies for other use cases with different requirements as well. In the summary, we give details which technologies fit our requirements given in Sect. 11.2.

Note that we focus on the technologies from an application layer perspective. There is a lot more related work regarding the lower layers of the OSI model. Additionally, there is lot more related work focusing on IoT (Internet of Things) scenarios with sensors instead of smartphones.

11.4.1   Bluetooth

In several related works published until 2011, data is exchanged using Classic Bluetooth and a direct device-to-device connection [26, 11, 29, 42]. In Spiderweb, each mobile device functions as server or client [29]. While being in the server role, the device publishes a service that clients can search for and connect to. On current devices and mobile OSs, Classic Bluetooth is still available. However, Apple restricts connections to other iOS devices, which makes device-to-device communication via Classic Bluetooth impossible for multiplatform applications (R1). The second requirement that Classic Bluetooth cannot fulfill is R2. For Classic Bluetooth data transfer, an explicit connection has to be established, which typically requires user interaction.[3]

There are several papers that propose solutions that avoid having to establish explicit connections between devices [33, 40, 9, 30, 36]. E-Shadow uses WiFi and Bluetooth without establishing explicit connections [33]. This is done by utilizing the WiFi SSID, the Bluetooth device name, and the Bluetooth Service Discovery Protocol (SDP) to broadcast data. The three different technologies are used to cover different physical ranges. The numbers given in [33] are as follows. For WiFi SSID, they indicate a range of 50 m and a size of 32 bytes. For Bluetooth device names, they state a range of 20 m and a size of 2000 bytes. For Bluetooth SDP, they state a range of 10 m and a size of 1000 bytes. Other related papers also suggest exploiting the Bluetooth device name [9] or Bluetooth SDP [20, 40, 34]. While the proposed Bluetooth-related solutions—device name and SDP—circumvent having to establish a connection, they have the drawback of being related to Classic Bluetooth. Thus, they do not benefit from the energy efficiency introduced with Bluetooth Low Energy (BLE). Furthermore, Apple's restrictions regarding the connection of Bluetooth devices in iOS within the MFi (Made for iPhone/iPod/iPad) program[4] could make it impossible to utilize these approaches.

---

[3]For example, in our prototype in Sect. 10.7, we used NFC for establishing the Bluetooth connection between two users.

[4]https://developer.apple.com/programs/mfi/.

In 2010, BLE was introduced in Version 4.0 of the Bluetooth Core Specification.[5] Many of the features of Classic Bluetooth are inherited, while a low latency and a very low energy consumption are achieved [41]. Most current smartphones support this standard.

Two different modes are available in BLE: undirected advertising/scanning (AD/SC) mode and central/peripheral connection (C/P) mode [41]. In AD/SC mode, small payloads can be broadcast without the need of an established connection. C/P mode allows larger payloads, but there has to be an explicit connection between devices. To be more specific, devices in peripheral mode can advertise their presence, while central nodes can connect to those nodes [41]. BLE is often used in IoT scenarios with sensors. Having multiple sensors in peripheral mode allows a central node to periodically connect to the sensors. Peripheral nodes cannot communicate with other peripheral nodes.

The authors of [41] present a framework called BlueNet for IoT scenarios that allows the switching of roles in BLE. Sikora et al. use AD/SC mode for data exchange between two smartphones [32]. They report a range of about 40 m and a size of 37 bytes. The authors report that, at least for Android, the device switches automatically between advertising and scanning mode, and the software developer can influence the delay between switches but not define it in an exact manner.

11.4.2   WiFi

WiFi, IEEE 802.11 standard, has two modes: *infrastructure* and *ad hoc* mode. Infrastructure mode is the common mode of devices connecting to access points, whereas ad hoc mode allows for communication between devices directly.

The infrastructure mode is specifically *not* device-to-device communication. However, by creating a WiFi access point and letting other devices join the created network, effectively, device-to-device communication can be implemented in this mode. With ShAir, a middleware with this approach is presented in [10]. Furthermore, popular file sharing applications use this approach, for example, the app Xender.[6] According to the FAQs on the app's website, file sharing is done by being in the same WiFi network or by creating access points on the smartphone. SHAREit[7] works the same way. The disadvantage of this approach is that running in the background is not possible without disrupting the user's experience: while a software based on this approach is active, the user probably is not able to be connected to his/her usual WiFi network. This violates R3 (transferring data in the background).

---

[5]https://www.bluetooth.com/specifications/archived-specifications/.

[6]Previously mentioned in Sect. 8.2 with related publication [39].

[7]https://www.ushareit.com/.

Some papers suggest using the WiFi SSID for broadcasting small amounts of data [33, 36] (cf. Sect. 11.4.1). However, iOS does not allow changing the WiFi SSID programmatically. Furthermore, if the SSID is bound to an opened access point on the device, this typically means that the device cannot be connected to another WiFi network, which again violates R3.

The main issue with WiFi ad hoc stems from the fact that it never was widely deployed in the market [7]. In [19], the authors use the WiFi ad hoc mode on an Android device. This mode is not available by default; an extension had to be compiled into the Linux kernel. There is still a lack of support of WiFi ad hoc since the publication of that paper (2016), which shows that only a very limited set of Android devices would be able to run such an application. Furthermore, WiFi ad hoc mode is not supported on iOS [36].

The WiFi Alliance developed another WiFi mode for device-to-device communication, WiFi Direct, which is supported on Android devices with version 4.0 and higher [25].[8] There are several related papers dealing with WiFi Direct for device-to-device communication [7, 38, 21, 2, 31, 17]. While the Talk2Me prototype was developed utilizing WiFi Direct, Shu et al. describe how it is not mature enough and wasn't used for the evaluation [31]. Instead they used UDP over WiFi with devices connected to the same access point. In WiFi Direct, every connection needs to be confirmed manually; workarounds might be possible though. By default, however, this violates R2 (no explicit connection establishment). Furthermore, iOS does not support WiFi Direct, violating R1 (multiplatform app). Apple instead offers its own device-to-device framework that is only available for iOS devices.

WiFi Aware, sometimes called NAN (Neighbor Awareness Networking) is another approach by the WiFi Alliance for device-to-device communication. Android implements its functionality with version 8. The developer websites indicate that its functionality is dependent on the actual WiFi hardware and firmware.[9] Although some websites report that WiFi Aware is based on Apple's AWDL (Apple Wireless Direct Link) technology, Apple does not seem to support WiFi Aware. According to the website of the WiFi Alliance, there are currently less than 50 smartphones specifically certified for WiFi Aware.[10] In the TYDR user base, being a small app for research purposes, we already see more than 500 distinct devices.

WLAN-Opp was developed based on IEEE 802.11 and tethering between smartphones [35]. It is supposed to serve as an alternative for WiFi ad hoc and WiFi Direct given their shortcomings and limited availability. The implementation of WLAN-Opp is for Android only and not maintained.[11]

---

[8]WiFi Direct has initially been called WiFi Peer-to-Peer.

[9]https://developer.android.com/guide/topics/connectivity/wifi-aware.

[10]https://www.wi-fi.org/product-finder-results?sort_by=certified&sort_order=desc& certifications=56.

[11]https://github.com/saschat/WLAN-Opp.

11.4.3 Frameworks

The open-source community and some companies have developed frameworks aiming to provide abstractions for device-to-device communication. Google and Apple as the major mobile OS providers also provide their own solutions. In this section, we will give an overview and highlight key characteristics.

In Sect. 8.2, we already mentioned AllJoyn, a framework allowing devices to communicate with other devices in proximity. There are a few projects building on top of AllJoyn [37, 18, 16]. As the latest release is from 2017, we assume the project is not actively maintained anymore.

Thali is an open-source project with the goal of enabling device-to-device computing. The code is not actively maintained[12] and only exists as a Cordova[13] plugin. The developers specifically highlight the issue of connecting Android and iOS in a device-to-device manner, stating they only found a workaround.[14] It consists of using BLE for finding other devices and then manually joining a WiFi access point opened on another device. This procedure violates R2 as manual user interaction is required.

Some companies offer frameworks for device-to-device communication. Uepaa AG's p2pkit[15] is a multiplatform framework for seamless device-to-device computing. However, the code does not seem actively maintained.[16] Open Garden's FireChat app[17] for offline messaging via message exchange in a device-to-device manner gained some attention during the times of government censorship and unavailability of Internet connections.[18] Open Garden's MeshKit SDK though, mentioned, for example, in [27], cannot be found online and is not part of the company's GitHub repository.[19] Bridgefy[20] follows the same goal of offline device-to-device communication. Their free plan allows for 30 monthly offline users.[21] Broadcasting, i.e., the connectionless sending of messages to devices in proximity, only works in the mesh mode of the framework, and the maximum message size then is 2048 bytes.[22]

There were and are some products and apps available with device-to-device functionalities. Hand-held gaming devices from Nintendo and Sony were offering

---

[12]https://github.com/thaliproject/Thali_CordovaPlugin.

[13]A framework for multiplatform mobile application development, see https://cordova.apache.org/.

[14]http://thaliproject.org/Android-and-iOS-interop/.

[15]http://p2pkit.io/.

[16]https://github.com/Uepaa-AG.

[17]https://www.opengarden.com/firechat/.

[18]https://en.wikipedia.org/wiki/FireChat.

[19]https://github.com/opengarden.

[20]https://www.bridgefy.me/.

[21]https://www.bridgefy.me/pricing.html.

[22]https://github.com/bridgefy/bridgefy-ios-developer/blob/master/README.md.

data exchange with nearby players in proximity.[23] This is a feature specific to each gaming system and does not work across devices from Nintendo and Sony.

Both Apple and Google provide frameworks that enable developers to build apps that are able to communicate with nearby devices. Apple's framework is called MultipeerConnectivity[24] and uses different technologies like WiFi and Bluetooth for communication and is only supported by iOS devices. Google's Nearby framework[25] uses technologies such as Bluetooth, WiFi, and audio and consists of two different APIs: Nearby Connections and Nearby Messages. Nearby Connections allows direct communication with other devices in proximity without the need for an Internet connection. It is only available for Android. Nearby Messages requires an Internet connection and only allows the exchange of small payloads, but it is available for both Android and iOS. Google describes that they utilize "a combination of Bluetooth, Bluetooth Low Energy, Wi-Fi and near-ultrasonic audio."[26] Using these technologies, tokens are exchanged between devices. After receiving a common token, Google's servers distribute the payload to the receiving device. Although the messages are relayed through Google's servers, the documentation emphasizes that "Nearby Messages is unauthenticated and does not require a Google Account."[27] The maximum payload size is 100 kibibyte, i.e., 102400 bytes.

11.4.4   Summary

Table 11.1 gives an overview of the device-to-device communication approaches that we disregard for our proposed system.[28]

Because of Apple's mentioned restrictions regarding certifications for Bluetooth devices, the three approaches related to Classic Bluetooth are not readily available on iOS: Classic Bluetooth, Bluetooth Device Name, and Bluetooth SDP. WiFi ad hoc, WiFi Direct, WiFi Opp, WiFi Aware, and Google Nearby Connections are not available on/for iOS devices. Apple Peer Connectivity is not available on Android devices.

Regarding R2, exchanging data without manual user interaction, we observed that iOS only allows the connection to new WiFi access points after manual user interaction. This leads us to disregard WiFi infrastructure mode and Thali which uses this workaround for device-to-device communication. Changing the WiFi SSID

---

[23] https://www.nintendo.com/3ds/built-in-software/streetpass/how-it-works   and   http://us.playstation.com/psvita/apps/psvita-app-near.html.

[24] https://developer.apple.com/documentation/multipeerconnectivity.

[25] https://developers.google.com/nearby/.

[26] https://developers.google.com/nearby/messages/overview.

[27] https://developers.google.com/nearby/messages/overview.

[28] Additionally, we disregard the solutions by Nintendo and Sony because they are proprietary solutions for their respective gaming devices and are not available for smartphones.

Table 11.1 Excluded approaches for device-to-device communication

Reason	Technology	Comment
R1	Classic Bluetooth	No iOS support
R1	Bluetooth Device Name	No iOS support
R1	Bluetooth Service Discovery Protocol (SDP)	No iOS support
R1	WiFi ad hoc	No iOS support
R1	WiFi Direct	No iOS support
R1	WiFi Aware	No iOS support
R1	WiFi Opp	No iOS support
R1	Google Nearby Connections	No iOS support
R1	Apple MultipeerConnectivity	No Android support
R2	WiFi infrastructure	No automatic connection on iOS devices
R2	WiFi SSID	Not programmatically changeable on iOS
R2	Thali	No automatic connection on iOS devices
R2	BLE C/P	Explicit connection establishment
R3	p2pkit	No background data exchange support for iOS to iOS and code not maintained
Other	Open Garden MeshKit	SDK not available
Other	AllJoyn	Code not maintained

is not programmatically possible on iOS, which violates R2 as well. BLE C/P needs the explicit connection between devices, so it violates R2 as well.

We note that p2pkit violates R3, the transfer of data in the background. It does not enable the exchange of data between two iOS devices that are not actively used.[29] The Open Garden MeshKit is not to be found and thus we exclude it. Both AllJoyn's and p2pkit's code does not seem to be maintained; the latest releases of both are 3 years old at the time of writing.

Implementing multiple device-to-device approaches in one app would likely result in interferences at the wireless interfaces or excessive battery drain. Both Apple MultipeerConnectivity and Google Nearby utilize BLE, for example. Using both technologies and trying to combine their capabilities this way would likely not work well because the BLE interface could most likely just be used by one of the frameworks at each time.

This leaves three options that fulfill R1, R2, and R3: BLE SC/AD, Bridgefy, and Google Nearby Messages. In the next section, we will design our application based on the results of this section.

---

[29]http://p2pkit.io/developer/support/faq/.

11.5   Design of MobRec

In Fig. 11.1, we illustrate the proposed general modular architecture of MobRec. The three main components of the system are *data collection, data exchange,* and *recommender system.* Data collection is responsible for getting data about the user. Data exchange is responsible for getting data from other users. The recommender system utilizes all available data for recommending items to the user. The mobile OS provides components for sensors (e.g., for tracking the user's location for inferring his/her favorite POIs) and wireless interfaces (for exchanging data).

External service providers might be needed (or be useful) in order to retrieve metadata about items, utilize existing systems, or offload data or computational tasks. Figure 11.1 shows dashed lines for optional connections to third-party service providers. Data collection might use this to retrieve data about the user or to enrich already available data, e.g., find out the genre of the songs the user listened to. The recommender system can optionally be relayed to an external service provider.

The system should be developed in a modular way in order to be able to exchange components easily. Consider the multitude of device-to-device approaches. Technological advances or the development of new frameworks could offer shorter connection times, and higher bandwidths, or larger transmission ranges. We then might want to exchange the data exchange module. Similarly, advances in recommender systems and machine learning might offer better recommendations, creating

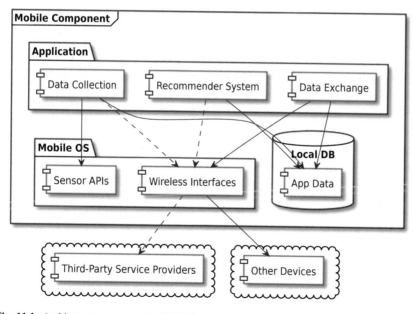

Fig. 11.1 Architecture components of MobRec

the need to replace the module or offload certain tasks to components available from external service providers.

11.5.1   Data Collection

We identify three different possibilities to retrieve user data:

Automatic Data Tracking The first part of this thesis covered the automatic tracking of context data (cf. Part I). In Chaps. 9 and 10, we were highlighting links between the user's behavior and user's characteristics and preferences. Additionally, papers like [15] show further links between behavior and implicit ratings. In [15], links are shown between geolocation histories and implicit place ratings. Thus, we assume to be able to use automatically tracked data for either finding similar users or finding implicit ratings.

In Part I, we were focusing on the data that can be collected on Android devices. The data that can be tracked automatically on iOS might differ. In order to create a multiplatform system and ensure that the same data points are available on all systems, additional ways of retrieving the user's ratings are necessary. Figure 11.2a shows the sequence diagram of automatic context data tracking (mobile sensing).

Querying Third-Party Service Providers In order to minimize necessary user effort, the second method we suggest is retrieving data from existing service providers. For example, Spotify's API enables application developers to fetch recently played tracks.[30] Similarly, both Apple Music[31] and Deezer[32] also allow developers to get most recently played tracks. According to Statista, these three music streaming service providers make up 57% of the worldwide music streaming market, with Spotify and Apple Music being the two biggest service providers.[33] Regularly retrieving recently played back music yields a complete music listening history indicating implicit user ratings. Figure 11.2b shows the sequence diagram for the collection of data from a third-party service provider.

Manual User Input For data that is neither automatically trackable nor available via third parties, the user should be able to enter it manually. By defining an ontology for categories and terms that can be exchanged between users, compatibility between the data from different collection methods can be ensured. Predefined categories can be movies, music, or restaurants, where recommender system are often used, but any other category would be possible as well. Service providers

---

[30] https://developer.spotify.com/documentation/web-api/reference/player/get-recently-played/.

[31] https://developer.apple.com/documentation/applemusicapi.

[32] https://developers.deezer.com/api.

[33] https://www.statista.com/statistics/653926/music-streaming-service-subscriber-share/.

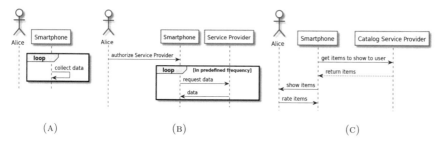

Fig. 11.2 Data collection mechanisms. (**a**) Automatic data tracking (mobile sensing). (**b**) Querying third-party service providers. (**c**) Manual user input

like The Movie DB (TMDB),[34] for example, can be used to help employ globally valid identifiers for each item, in this case, for each movie. Figure 11.2c shows the sequence diagram for manual data collection. *Catalog Service Provider* denotes a service provider that offers structured information about a specific category, like the mentioned The Movie DB.

11.5.2   *Data Exchange (Device-to-Device Communication)*

The three remaining approaches for device-to-device communication from our overview in Sect. 11.4 are BLE SC/AD, Bridgefy, and Google Nearby Messages. All of them seem to fulfill R1, R2, and R3. None of them, however, fulfills R4, transferring larger payloads. In this section, we present our workaround utilizing cloud storage providers. We then investigate the three remaining technologies to decide which one to choose for the implementation.

Size Limitation Workaround with Cloud Storage Providers Building on the existing device-to-device approaches, we present a workaround to facilitate the broadcasting of large payloads while fulfilling R1 (multiplatform app), R2 (no explicit connection establishment), and R3 (data transfer in the background). Figure 11.3 visualizes our workaround. First, Alice authorizes the system to access her account at some cloud storage provider (CSP) like Dropbox, Google Drive, etc. Alternatively, she could use her own cloud storage. In some predefined frequency, Alice's data is then uploaded to the CSP and shared via a public URL. This URL is then broadcast via one of the abovementioned approaches. As only the URL is shared, which can be further shortened via a URL shortener service, the available small payloads should suffice. Another user, Bob in Fig. 11.3, receives the broadcast with the URL and can download Alice's publicly shared data. Optimizations like

---

[34]https://www.themoviedb.org/.

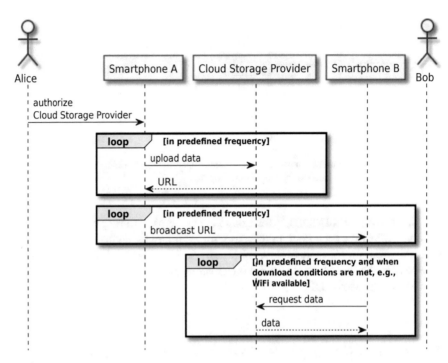

Fig. 11.3 Design of the data exchange via a third-party cloud storage provider

waiting for a WiFi connection can easily be implemented. Note that the only required user interaction by Alice or Bob is the authorization of the CSP, which only has to be done once.

Bluetooth Low Energy (BLE) When using BLE AD/SC mode, custom data can be sent in different fields that are part of the advertisement data. In that advertisement data, we could broadcast the URL pointing to Alice's data at her CSP. In order to broadcast, the device needs to be in peripheral mode, which both Android and iOS support, fulfilling R1. Both scanning and advertising do not require manual user interaction, fulfilling R2. Android allows for both scanning and advertising while the app is in the foreground or background, fulfilling R3. Apple also allows Bluetooth-related tasks to be done while running in the background. However, the scanning intervals are longer which might lead to two passing users missing each other if they do not stay in proximity for long enough. We conducted tests on real devices that showed that iPhones advertising while our app was running in the background could not be discovered by any other device (tested with Android smartphones, iPhones, and macOS laptop). Because of this limitation, upon closer inspection, we do not consider R3 fulfilled.

Google Nearby Messages R1 and R2 are fulfilled for Google Nearby Messages. Looking deeper into R3, exchanging data while the app is in the background, for

Android, the documentation describes that scanning should only be done while in the foreground. However, in the background, it is still possible to scan for beacon messages.[35] In order to do that, the Nearby Messages client needs to specify a strategy that only uses BLE. For iOS, both background advertising and scanning are supported. Again, a strategy only using BLE has to be defined for this. It seems like Google worked around the issues regarding iOS and background advertising we reported about in the previous paragraph. We did not find an exact description how Google implemented this. Google's documentation states that background subscriptions are more energy-efficient but provide lower reliability and higher latencies. The remainder of Chap. 11 will show that Google Nearby Messages is still a viable solution despite its limitations.

Bridgefy Comparing Bridgefy to Google Nearby Messages, we see two major drawbacks for Bridgefy. First, the maximum payload size is 2048 bytes, whereas the payloads in Nearby Messages can be 50 times that size. The bigger payloads in Nearby Messages will allow us to send more data before downloading data from the cloud storage provider. Second, the framework is a commercial product, and the free version restricts the number of offline users. Thus, while Bridgefy in principal might be a viable solution, we opt to go with Google Nearby Messages for our prototype.

Summary The data transfer with Google Nearby Messages is, because of the described relay over Google's servers, strictly speaking not direct device-to-device transfer between two devices. We still chose it for our prototypical implementation because of the described benefits of being free of cost and supporting a larger payload. In our view, the ubiquity of Internet connections allows for using a service that requires Internet connection. Because of the modular design, we could replace the data exchange module with one utilizing Bridgefy and then having direct device-to-device communication.

11.5.3   *Recommending New Items*

This chapter focuses on an architecture that facilitates decentralized recommender systems. In the prototype, we will relay the recommendation task to external service providers. In this section, we sketch the challenges recommender algorithms in MobRec will face and pose potential solutions.

When employing a local recommender system on the smartphone, additional data is needed. For content-based filtering, the properties of items have to be known. Third-party service providers can help with retrieving such needed metadata about items. For user-based collaborative filtering, information about the similarity of users is utilized. Whereas services like Spotify or Netflix have very large databases

---

[35]https://developers.google.com/nearby/messages/android/get-beacon-messages.

with millions of users, the local databases in MobRec will be much smaller, and thus there is a lower likelihood of finding similar users.

We see two possible solutions for this problem. First, we could let each user disseminate more than just his/her own item preferences/ratings and let him/her also send data from previous encounters—this would also address the cold start problem new users will face. Another approach is to calculate the similarity of users in a different way, independent of the users' ratings. In Sect. 8.1, we described the propinquity effect of physical proximity being a good predictor of forming interpersonal bonds. Having a unique identifier for each user and counting the number of times and/or the duration of being in proximity would then likely predict a higher bond. Additional methods are available for determining similarity in proximity-based applications. In Chap. 10, we developed and evaluated a method for estimating similarity based on the exchange of the users' context data using probabilistic data structures. In [13], we developed a privacy-preserving method for determining the similarity of two users based on their text messaging data. Both of those methods can be implemented in our proposed architecture to find similar users, without having the need to have users who rated the same items. Future work will have to show to what extent the propinquity effect or the mentioned similarity metrics yield valuable similarity indications for user-based collaborative filtering. Future work could also investigate the feasibility of approaches like federated learning, effectively exchanging trained models or updates to models for recommendations [22].

Following this idea of having separate similarity data and ratings, in the following sections, we distinguish between two data types:

- *simdata*
- *ratingsdata*

The basis is the assumption that an estimated similarity, for example, based on smartphone data, will yield an indication of similar ratings of items. Thus, not only when we find similar users via *ratingsdata* but also when finding similar users via *simdata* can we recommend new items to the user. Note that we do not evaluate this assumption because that would likely require the deployment of the whole system and data collection with lots of users including feedback on the given recommendations. Instead, our implementation uses existing third-party service providers for recommendations based on similar people (determined by *simdata*) that the user has met.

In order to keep the information fresh, MobRec can simply (re-)download data from users met in the past. The process is then that (at least some) data from each user is automatically tracked, either by mobile sensing or from external service providers, and updated on the user's cloud storage provider. After Bob has met Alice, he knows her URL and can just download her latest data. In our prototypical implementation, where recommendations are relayed to external service providers, up-to-date data is available, for example, TMDB is updated constantly, and the latest movies can be recommended.

Recommending Items for Groups of Users Another field which has gained less attention in industry and academia is that of group recommender systems, e.g., [8]. With its ad hoc nature and immediate preference data exchange, MobRec is ideally suited to be used for pervasive group recommendation scenarios. Exchanging data between several users in a group setting, a local recommender system can calculate recommendations based on the given data, considering the preferences of each user. When utilizing an external service provider for a recommendation, most likely, before contacting it, the preferences of each group member have to be combined into one group profile as most providers will only recommend items for a single user. We follow this approach in Chap. 12.

11.6 Implementation of MobRec

In this section, we describe the implementation of MobRec. The idea is to have a minimum viable product (MVP) that shows all core functionalities. In the following, we present the core modules of our architecture and describe what frameworks and third-party service providers we utilized.[36]

11.6.1 Multiplatform Development

In order to be able to reach almost 100% of all smartphone users, an app for Android and iOS has to be developed (R1). For the implementation of our MVP, we opted for Ionic,[37] which is an SDK (Software Development Kit) built on top of Cordova, a framework for multiplatform development. Using such multiplatform frameworks is an alternative to developing two distinct apps, allowing to have one code base for both apps. Multiplatform frameworks take a few different approaches in how they work. Often, the differences between the approaches lie in the programming language used and in how the UI is rendered. The latter often either is part of a web component that is displayed within a browser inside the app or is rendered with native components. This typically results in a trade-off between performance (native is better) and and ease/speed of development (webapp is faster). Looking at statistics about the most used frameworks among developers from 2019,[38] for multiplatform development, we observe that of the surveyed developers, 10.5% reported using React Native, and for Cordova it is 7.1%, for Xamarin 6.5%, and for Flutter 3.4%.

---

[36]In this section, we focus on the principles of the implementation. For a detailed description about the implementation of each module, see [12].

[37]https://ionicframework.com/.

[38]https://www.statista.com/statistics/793840/worldwide-developer-survey-most-used-frameworks/.

11.6.2   *Data Collection*

In this section, we give details about the implementation of the data collection in MobRec, structured by the three methods given in Sect. 11.5.1.

Automatic Data Tracking Cordova plugins for accessing the user's location, also while the app is running in the background, are readily available.[39] When tracking the user's location, frequent visits at points of interest or restaurants can indicate preference, and ratings can be inferred. When implementing location tracking, the trade-off is typically between accuracy, frequency, and battery drain. For the users, no interaction is required besides the system confirmation that our app can access his/her location.

   In our implemented MVP, we use the location traces of a user as *simdata*. Each location point is first transformed into a Geohash,[40] a short string representation of a latitude/longitude pair. Then, each of the user's locations are entered into a one-hash CBF and compared via CBF-Dice (Eq. 10.6; see Chap. 10).

Querying Third-Party Service Providers Some third-party service providers allow the user—or an application on behalf of the user—to export the items consumed with that provider. This is an easy way to track the user's taste. Listening to music is one of the most common activities with smartphones.[41] Spotify has by far the most subscribers in the market of music streaming services (36% market share.[42]) Given the user's permission, we access the user's 50 most recently played tracks using OAuth 2.0.[43] Retrieving these recently played tracks regularly yields implicit ratings by the user—based on the assumption that the more a user listened to a track, the more he/she likes it. From the user's side, authorizing our app to access Spotify is the only action he/she must take.

Manual User Input Globally, 37% of Internet users use Netflix.[44] Watching movies and TV shows is a common pastime, and recommending new items in these fields is a common task for recommender systems. At the time of writing, Netflix does not offer a publicly available API, though their website offers the functionality of downloading a *viewing activity* list. However, if our system wants to recommend movies, only tracking those movies available on Netflix will limit the available range of movies: as of 2018, Netflix only offered 4010 movies (in the USA).[45]

---

[39]https://www.npmjs.com/package/@mauron85/cordova-plugin-background-geolocation.

[40]https://web.archive.org/web/20080305223755/http://blog.labix.org/#post-85 and http://geohash.org/.

[41]https://www.pewresearch.org/internet/2015/04/01/us-smartphone-use-in-2015/.

[42]https://www.statista.com/statistics/653926/music-streaming-service-subscriber-share/.

[43]https://developer.spotify.com/documentation/web-api/reference-beta/#endpoint-get-recently-played.

[44]https://www.statista.com/statistics/758369/netflix-video-usage-region/.

[45]https://www.businessinsider.de/netflix-movie-catalog-size-has-gone-down-since-2010-2018-2.

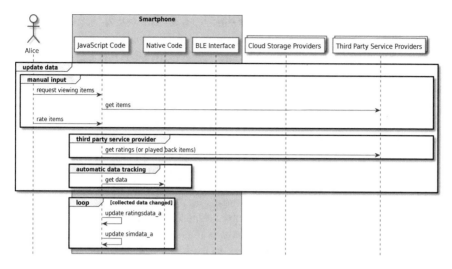

Fig. 11.4 Data collection in MobRec

With around 650 films released each year in the USA alone,[46] this is not a high number. We use the publicly available *The Movie DB*[47] (TMDB) API to create a visual interface for the user to rate movies. The Movie DB contains 534,375 movies.[48] When designing the visual interface and functionality, we followed the approach of the MovieLens project as described in [24]. This includes searching, rating movies, adding them to a watchlist, and feedback on recommendations (rate, add to watchlist, not interested). Additionally to movies, TV shows are also available via the TMDB API and thus available for the users of our prototype to rate.

Figure 11.4 shows all three approaches for collecting and processing data. Note that only the manual user input needs user interaction. Getting data from third-party service provider Spotify only requires a one-time authorization (not shown in the figure). Automatic data tracking is done in the background and uses native libraries of Android and iOS. Once the collected data changed, we can automatically update Alice's *ratingsdata* and *simdata*, denoted in the figure as belonging to Alice with the suffix *_a*.

---

[46] According to http://data.uis.unesco.org/.

[47] https://www.themoviedb.org/.

[48] https://www.themoviedb.org/faq/general; accessed 2020-03-30.

11.6.3   Data Exchange (Device-to-Device Communication)

As described in Sect. 11.5, the data exchange with nearby users is designed to utilize the Google Nearby Messages API. As the library only supports small payloads of 100 kibibyte, the workaround with uploading the user's *ratingsdata* and *simdata* to a CSP and sharing the public URL of that file was used. In the following, we present details about the utilization of the Google Nearby Messages API and the sequence of the data exchange.

The Google Nearby Messages API for Android is available in Java and Kotlin, and iOS developers can use Swift or Objective-C. In order to integrate the library into an Ionic application, a plugin is required to invoke calls to the native libraries from the JavaScript code. During development in the context of a master's thesis, we only found one Cordova plugin that supports the Google Nearby Messages API[49] [12]. However, the implementation is only available for Android. Furthermore, the Android implementation is not configured to work in the background. Therefore, as part of the master's thesis, a custom plugin was designed and implemented.[50] Both on Android and on iOS, the publishing strategies are set to work in the background, utilizing BLE.

Based on using the described workaround for device-to-device size limitations (cf. Sect. 11.5.2), additionally to *ratingsdata* and *simdata*, we define the following data types:

- *dataset.* This contains the *ratingsdata* and optionally additional information like a nickname or profile picture, etc. Note that it does not contain *simdata*.
- *cspurl.* This is the URL pointing to the publicly available *dataset* at a cloud storage provider (CSP).

Figure 11.5 shows the initialization of the device-to-device communication utilizing the workaround with a CSP. Alice authorizes access to her account with the CSP. In the MobRec MVP, we use Google Drive. The platform-independent JavaScript code then handles the authorization for Google Drive via OAuth 2.0 and uploads Alice's dataset *dataset_a*. Note that if some data types are not present, for example, because Alice did not rate any items yet, parts or the whole set might be empty. The CSP returns *cspurl_a*.

Figure 11.6 shows the sequences for updating data at the CSP, broadcasting and scanning via the BLE interface, and receiving broadcast messages. Whenever *ratingsdata_a*, the nickname, profile picture, etc., changes, *dataset_a* is updated. *cspurl_a* stays the same and does not need to be updated. In our implementation, *simdata_a* is small enough to be sent with the payload broadcast via Google Nearby Messages. We trigger the broadcasting of messages from the JavaScript code. The publishing itself, i.e., broadcasting messages via Google Nearby Messages via BLE,

---

[49]https://github.com/hahahannes/cordova-plugin-google-nearby.

[50]The plugin was developed by Simone Egger as part of her master's thesis; for details, see [12].

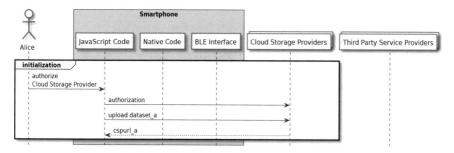

Fig. 11.5 Initialization of the device-to-device communication

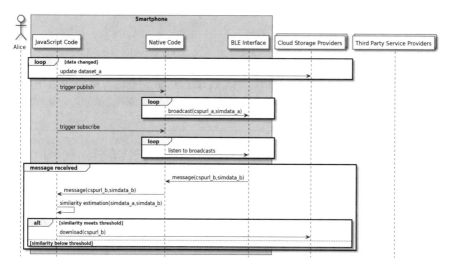

Fig. 11.6 Device-to-device communication

is done with our native code plugin. The broadcast message consists of the *cspurl_a* and *simdata_a*. Similarly, subscribing, i.e., listening for broadcast messages from other app instances in BLE range, is also triggered via JavaScript code and executed via our native code module.

Once a message, i.e., broadcast, is received, its content is handed to the platform-independent JavaScript code and processed there. First, the received *simdata_b* is compared to the phone's user's *simdata_a*. If the similarity comparison meets a predefined threshold, *dataset_b* is downloaded via *cspurl_b*. This means that we avoid downloading data from other users if the predefined similarity threshold is not met (cf. bottom of Fig. 11.6).

11.6.4   Recommendations

Instead of implementing our own recommender systems, for the MVP, we used external third-party service providers. For music recommendations, we utilize Spotify. Their API can return music recommendations based on up to five so-called seed tracks entered.[51] For movie recommendations, we utilize the TMDB API. Given a movie or TV show, other items are recommended.[52] Based on these APIs, we recommend new items to Alice based on Alice's own preferences and based on the preferences of similar people that Alice met.

11.6.5   Minimum Viable Product

In the course of her master's thesis, Simone Egger implemented the MVP [12]. Figure 11.7 shows screenshots. Figure 11.7a shows music recommendations based on the user's own listening history, and Fig. 11.7b shows music recommendations

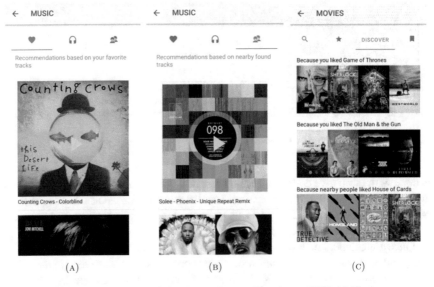

(A)   (B)   (C)

Fig. 11.7 Recommendations via third-party service providers in our MVP. (**a**) Music recommendations based on the user's favorite tracks. (**b**) Music recommendations based on people met. (**c**) Movie and TV show recommendations based on user's favorites and based on people met

---

[51] https://developer.spotify.com/documentation/web-api/reference/browse/get-recommendations/.

[52] https://developers.themoviedb.org/3/movies/get-movie-recommendations   and   https://developers.themoviedb.org/3/tv/get-tv-recommendations.

based on similar users met in proximity. Figure 11.7c shows how movie recommendations are displayed. Each row indicates to the user why the recommendations are being displayed, some based on own preferences, some based on users previously met.

11.7 Evaluation

In this section, we will evaluate our system, focusing on data exchange via device-to-device communication. Based on the concept of users exchanging data with other users, we analyze different scenarios in order to develop a concept of how to conduct the evaluation. During the design of MobRec, we already accounted for requirements R1–R4. There is no user interaction necessary for data exchange (R2), and we worked around limitations on payload size (R4). In this section, we investigate how well the data exchange works in the background (R3) and if there are any differences observable for Android and iOS (R1).[53]

Imagining the average users, most time is probably spent at home or at work. In those cases, the distance to other users will be very short, and the time spent in proximity is rather long, be it during a meeting or while sleeping. Furthermore, chargers will likely be ubiquitously available. Physical distance, the time it takes to discover nearby devices and exchange data, and battery consumption thus are not critical in this scenario. At busy workplaces, there could be interferences if there are a multitude of devices present though. We assume that Internet connectivity, required in order for Google Nearby Messages to work, is available in almost all home and work scenarios. In order for our concept to work properly, users need to meet new people to exchange data with though. Home and work location will thus not be the crucial situations where users exchange data.

Another scenario is to spend time together at some public or private place. This could be some event like a restaurant visit, a music show, or any other leisure time activity. Here, the time window might be shorter than at home or at work but is probably still at least around 1 h. The distance between users probably ranges from a few to around 50 m. Depending on the location, Internet connectivity might not be as good as in the previous scenario.

The third scenario is just passing other users, for example, when commuting via public transport. The time window of being in proximity might be rather short, e.g., waiting for the metro for a few minutes. We assume the distance to be short, from a few to around 20 m. Internet connectivity might be bad or at worst non-existent.

We also made tests regarding the physical distance between devices. As our system uses Google Nearby Messages with BLE, we assumed the distances between

---

[53]The experiments for this evaluation were executed by Simone Egger as part of her master's thesis [12].

devices to be unproblematic. We confirmed this with tests in both indoor and outdoor situations. Details about the distance tests are omitted here.

This leaves the following aspects to consider, which we cover in the following sections:

- **Multiple devices** We will check whether it is feasible to exchange data with multiple devices and whether the presence of multiple other devices has negative effects on the data exchange.
- **Discovery time** The time needed for successful discovery of present devices in proximity and data transfer. The transfer here just refers to receiving the braodcast data, *simdata* and *cspurl*, as downloading the *dataset* from the CSP is not time-critical and can be done later.
- **Internet connectivity** We will analyze to what extent bad Internet connectivity influences the data exchange.

In Sect. 11.7.4, we will summarize the evaluation and propose optimizations for battery consumption.

11.7.1 Multiple Devices

In our test setup, we used two iOS devices (both of them iPhone 6, iOS 12.2; in the following, we distinguish the two devices with (a) and (b)) and two Android devices (Xiaomi Mi A2 with Android 9 and LG K8 with Android 7). All devices were placed next to each other and started broadcasting and scanning at the same time. We recorded the timestamps for the start of broadcasting and for receiving the messages from the other devices. The test was conducted in a busy restaurant. This way, we simultaneously tested the feasibility of sending/receiving data from multiple devices at the same time and potential interferences by other nearby devices. We repeated the test five times. The results of the tests are shown in Table 11.2. The time given in the table is the time between the start of broadcasting/scanning and receiving all three messages from the other devices. Additionally, we show the average time. The LG K8 was the slowest to receive all messages in all test runs. However, the maximum was only 6.2 s. Overall, this test indicates that even with multiple devices and in busy places, all messages are received reliably in a matter of seconds. While the whole system might not scale indefinitely, we regard this test as evidence that the data exchange between multiple devices works well.

11.7.2 Discovery Time

One crucial factor for the evaluation of our system is the time it takes for devices to send and receive broadcast messages, i.e., finding nearby devices and receiving

Table 11.2 Time in seconds until devices received messages from all other devices. Tests conducted in busy restaurant

Test run	LG K8	Xiaomi Mi A2	iPhone 6 (a)	iPhone 6 (b)
1	3.00 s	1.40 s	1.94 s	0.06 s
2	6.20 s	2.00 s	1.00 s	1.00 s
3	4.30 s	2.60 s	1.40 s	1.34 s
4	5.70 s	1.50 s	1.03 s	0.09 s
5	2.30 s	2.10 s	2.10 s	1.10 s
Average	4.30 s	1.92 s	1.49 s	0.72 s

Table 11.3 Experiments conducted regarding devices finding each other. Shows the five possible combinations C1–C5

Combination	Proximity	App start	Device known	Results
C1	in p.	Now	No	Table 11.2 l. 1
C2	in p.	Now	Yes	Table 11.2 l. 2–5
C3	in p.	Past	Yes	Table 11.4
C4	Moving into p.	Past	Yes	Same as C3
C5	Moving into p.	Past	No	Table 11.5

the Google Nearby Messages payload. In order to evaluate this, we consider three binary variables:

- The devices can already be *in proximity* or *move into proximity*.
- The app start can be *now* (started at the beginning of the experiment) or in the *past* (i.e., the app is already running for some time).
- The device to be found is already known or not; i.e., a broadcast message from that device was already received in the past or not.

Table 11.3 gives an overview of all possible combinations C1–C5. Three binary variables yield eight overall possible combinations. Three combinations are not possible: two devices *moving into proximity* cannot be combined with app start *now*. The app already has to be running when moving into proximity. This leaves out two cases (with device known *yes* and *no*). Additionally, when two devices are in proximity and the app start is in the past, then it is not possible that the devices do not know each other already.

The results for C1 and C2 are already given in Table 11.2. All devices find each other in a matter of seconds, regardless if the devices have received messages from each other before or not. In these experiments, the app's start was at each test run's start. The test from Sect. 11.7.1 indicates that if the app's start is *now*, discovery time is at most a few seconds.

With C3, we test the time between messages received from the same device. Here, the app's start lies in the past, and broadcasting and scanning runs continuously in the background. For this test, we used one Android device (Xiaomi Mi A2, Android 9) and one iOS device (iPhone 6, iOS 12.2). We assume that our test results are still generalizable, as the implementation only differs between different

Table 11.4 Average time between messages in minutes (C3)

	Android	iOS
Avg. of test run 1	47.13 min	8.00 min
Avg. of test run 2	29.92 min	8.19 min
Avg. of test run 3	57.77 min	11.19 min
Avg. of test run 4	49.23 min	9.20 min
Average	46.01 min	9.14 min

platforms, not different devices of the same platform. We let both devices broadcast and scan for test periods of 5 h and recorded when messages were received. Table 11.4 shows the average time between received messages for four test runs, as well as the average time between messages. On Android, the other device was found at least once per hour, whereas on iOS, the time between messages was around 10 min. Thus, if two devices already exchanged messages before, subsequent messages are received in a lower frequency.

C1, C2, and C3 consider scenarios where the devices are in proximity. In the following, we consider scenarios where two devices move into proximity. In this case, the app start always lies in the past. We distinguish between devices that already exchanged messages before (C4) and those that did not (C5).

In C4, the devices already exchanged messages before. We let the devices move into proximity of each other and recorded the time until messages where received. Repeating the test five times, we got roughly the same results as for C3. This indicates that when the app start lies in the past and the devices already exchanged messages before, it does not make a difference if the devices are already in proximity or move into proximity during runtime.

We note the significant difference between first message (app start *now*, C1/C2) and subsequent messages (app start *past* and device known *yes*, C3/C4)—few seconds vs. several minutes (iOS)/up to 1 h (Android). A possible reason for this could be that if two devices already have exchanged messages before, the token for the message was already exchanged and is not sent again until it is renewed. When restarting broadcasting/scanning, the token might be renewed, and thus, messages are received immediately on both sides after starting the scanning. The exact internal mechanisms of Google Nearby Messages are not public and we are not sure when exactly tokens are renewed. We assume that Android and iOS work differently, either regarding the token or regarding the BLE interface or implementation provided by the OS. This would explain the different results for the different platforms. A possible workaround for long time intervals between messages from the same devices could be to restart broadcasting/scanning in a predefined frequency or depending on some other factors like location changes.

The last case to evaluate is C5. It is the same as C4, only that the devices have not exchanged messages before. We performed the test five times. The results

Table 11.5 Time in minutes
until message is received after
moving into proximity (C5)

Test run	Android	iOS
1	4 min	4 min
2	10 min	7 min
3	2 min	2 min
4	1 min	1 min
5	2 min	2 min
Average	3.8 min	3.2 min

for C5 are shown in Table 11.5.[54] For both Android and iOS, all messages were received in a time of less than or equal to 10 min. On average, messages were received after approximately 3–4 min after devices were in proximity. This is a longer time compared to the results when the broadcasting/scanning was just started and messages were received after a few seconds (C1/C2). It is also significantly less compared to C3/C4. The results from C5 show that when broadcasting/scanning is already running in the background, messages are received after a longer time even though no messages have been exchanged before.

A possible reason for this could be that when restarting broadcasting/scanning, tokens are exchanged immediately for the first time and then only in a specific interval of around 1–10 min. Battery optimizations by the OS could lead to the inconsistent times for each test run. A possible workaround for this could also be restarting broadcasting/scanning.

11.7.3   Internet Connectivity

Our system utilizes Google Nearby Messages, which requires an Internet connection in order to facilitate the actual message exchange between devices. In this section, we evaluate to what extent this message exchange is influenced in situations where connectivity might be bad, e.g., inside of some underground metro stations.

In order to consistently and reproducibly simulate bad Internet connectivity, we used the iPhone's built-in "Network Link Conditioner." It can simulate different network conditions including "very bad network," which we used in this experiment. It constraints the speed to 1000 kilobyte per second and simulates a packet loss of 10%.

We let one device broadcast messages and then let the iPhone scan for messages while being constraint to "very bad network" conditions. We logged the time it took to receive a message. The experiment was repeated ten times. As a means of comparison, we repeated the same experiment with LTE connectivity.

---

[54]Note that we do not give seconds accuracy here in order to account for the small inaccuracies introduced by not measuring the time it took to move into proximity.

Table 11.6 Time in minutes until message is received with different network conditions

	Android		iOS	
Test run	Bad conn.	LTE	Bad conn.	LTE
1	0:40	0:02	2:00	0:01
2	1:45	0:02	0:10	0:01
3	0:06	0.02	1:00	0:01
4	2:00	0:02	0:47	0:01
5	0:52	0:02	3:12	0:01
6	0:41	0:02	9:41	0:01
7	0:57	0:02	0:33	0:01
8	1:31	0:02	0:04	0:01
9	0:14	0:02	0:02	0:01
10	0:14	0:02	0:04	0:01
Average	0:54	0:02	1:45	0:01

Android does not have a similar built-in feature to simulate network conditions. In order to perform the experiment under the same conditions as with the iPhone, we set up a WiFi hotspot on an iPhone, given the "very bad network" constraint, and let the Android device (Xiaomi Mi A2) connect to it. Here, again, we conducted ten test runs with both "very bad network" and LTE.

The results of the test are shown in Table 11.6. Note that with respect to the three binary variables introduced in Sect. 11.7.2, this is an experiment with combinations C1/C2. For bad network conditions, the time until a message is received is significantly higher. But still all messages in the test run were received with a maximum discovery time of 9:41 min. On average, each message was received almost instantly via LTE (confirming the results from Table 11.2). The average delivery time for bad connectivity was 1–2 min.

11.7.4   Evaluation Summary and Battery Drain Optimizations

We summarize the key results of our evaluation as follows:

- Messages from multiple devices in busy scenarios are sent and received without issues within seconds.
- (Re)starting the broadcast/scan mechanism makes the device receive message in a matter of seconds.
- New devices in proximity can be discovered in 3–4 min.
- Discovery of devices met before is slow—on average 9 min (iOS) and 46 min (Android).
- Bad Internet connectivity will introduce an overall negligible delay in discovery time of around 2 min.

Looking back at the scenarios we described for exchanging data between devices, most of them are realizable. The discovery time of new devices of 3–4 min might lead to some missed opportunities of data exchange in quickly moving scenarios like waiting at the metro station. The longest time window was between messages from the same device. In those cases, the user would receive the same data anyway, which would not help to improve the performance of the recommender systems. Even if we assume new data is present, there is a simple workaround: the data that is transferred is the *cspurl*, which does not change when the *dataset* is updated. We can just check if the *dataset* changed and re-download from the users previously met. This way, each user would have only to be met once. On the other hand, this reduces the recognition of meeting the same user multiple times—which could indicate similarity. Also, changes in *simdata* would be missed.

Regarding battery drain, permanently running broadcast/scan in the background accounts for roughly 5% (Android) to 10% (iOS) of battery consumption per hour. Such a battery drain is not acceptable for real-world deployment. However, significant improvements for both average discovery time and battery drain could be easy: restarting the broadcast/scan mechanism depending on specific times and locations will improve both aspects at once. Consider the following naive optimization. We assume that the time of each smartphone is running in sync, as they usually use online servers to sync their time. Then, we can let our app turn on broadcasting/scanning at the exact same time on every phone for 2 min. Our evaluation shows that 2 min is enough to reliably find most devices in proximity, even during bad network connectivity. We could let the app broadcast/scan for 2 min every 15 min, as long as there has been a location change. We assume that on average at least during 16 h of the day, there won't be location changes (sleep and work). This leaves 8 h, each of which has four 15-min intervals. Multiplied by 2 min of broadcast/scanning, this yields 64 min of running in the background instead of 24 h. This would reduce the battery drain to less than 5% of its original value and likely still produce a lot of the data exchanges that would happen during permanent broadcasting/scanning. While we have not tested this, the implementation of this optimization should be possible with Ionic's background mode.[55] If that fails and native code is necessary, in Android, the Alarm Manager[56] can fire events at exact times. In iOS, a workaround might be necessary, for example, by utilizing media playback to keep the app from being suspended.[57]

11.8   Summary

Current recommender systems often exhibit lock-in effects for the user. Recommendations might be biased according to the interests of the providing platform and

---

[55] https://ionicframework.com/docs/native/background-mode.

[56] https://developer.android.com/guide/background#alarmmanager.

[57] https://developer.apple.com/documentation/avfoundation/media_assets_playback_and_editing/creating_a_basic_video_player_ios_and_tvos/enabling_background_audio.

are often bound to the items available through the platform. We proposed a decentralized mobile architecture for recommender systems, MobRec, that leverages the preferences/ratings from users that are, or have been, in proximity. The introduced system runs on the users' smartphones and utilizes existing external third-party service providers. It is built on the general concept that similar people like similar things.

MobRec consists of three main modules: *data collection*, *data exchange*, and *recommender system*. We developed three ways to collect data: tracking context data automatically, retrieving data from third-party service providers, and letting the user manually rate items. We highlighted that while short-range wireless transmission technologies are implemented on all modern smartphones, exchanging larger amounts of data in the background without user interaction on a system available for both Android and iOS remains a challenging task. We proposed a workaround by broadcasting URLs of cloud storage providers from which the receivers can then download the sender's data.

We extensively reviewed the state-of-the-art of device-to-device communication and opted for Google Nearby Messages in our implementation. The evaluation of our MobRec prototype shows that the discovery time—the time needed to find other devices and exchange data—is just a few seconds when the broadcasting/scanning mechanism was just started. Overall, new devices in proximity are discovered within 3–4 min on average. Devices previously met are discovered again at a much slower rate, from around 10 min (iOS) to around 46 min (Android) on average. Because Google Nearby Messages requires an Internet connection, we also evaluated the influence of bad Internet connectivity and found that it introduces a delay of about 1–2 min on average. Battery drain remains an issue with constant broadcasting/scanning. We proposed simple optimizations that should reduce the battery drain by about 95%: only broadcasting/scanning during short fixed time intervals and only if the physical location changed.

Regarding the recommendation process, when considering user-based collaborative filtering, we looked into ways for determining the similarity between users, including methods independent of the users' ratings, for example, based on our previous work in Chap. 10. In order to have a fully functional minimum viable product without deploying our system with real users and collecting real data for locally calculated recommendations, we utilized external service providers for music and movie recommendations. The recommendations displayed are based on the user's own preferences as well as based on similar users' tastes.

Future work includes the deployment of the system with real users. Regarding the recommender system, future work includes the implementation of a mobile recommender engine operating on locally available data. A simulation with a real dataset, for example, from TYDR, could help evaluate the quality of the recommendations that such a system can provide. Furthermore, potential attack vectors and the overall security of the system should be investigated.

Besides being a general platform for recommender systems, MobRec is ideally suited for USN group scenarios: in an ad hoc manner, a group of users can use some device-to-device communication feature that exchanges data between the users in order to provide some service based on the shared data. In that case, R2 and R3, the broadcasting of data in the background without any user interaction, would not be applicable, making it possible to utilize a broader range of the device-to-device technologies presented in Sect. 11.4. In the following chapter, we follow the idea of providing USN services for groups of users and present a music recommender system.

References

1. M. Ahmed, Y. Li, M. Waqas, M. Sheraz, D. Jin, Z. Han, A survey on socially aware device-to-device communications, in *IEEE Communications Surveys Tutorials* , vol. 20(3) (2018), pp. 2169–2197. https://doi.org/10.1109/COMST.2018.2820069
2. N. Aneja, S. Gambhir, *Profile-Based Ad Hoc Social Networking Using Wi-Fi Direct on the Top of Android* (2018). https://doi.org/10.1155/2018/9469536
3. R. Baraglia, P. Dazzi, M. Mordacchini, L. Ricci, A Peer-to-Peer recommender system for self-emerging user communities based on gossip overlays, in *Proceedings of the 10th IEEE International Conference on Computer and Information Technology.* Journal of Computer and System Sciences, vol. 79 (2) (2013), pp. 291–308. https://doi.org/10.1016/j.jcss.2012.05.011
4. L.N. Barbosa, J. Gemmell, M. Horvath, T. Heimfarth, Distributed user-based collaborative filtering on an opportunistic network, in *Proceedings of the 2018 IEEE 32nd International Conference on Advanced Information Networking and Applications (AINA)* (2018), pp. 266–273. https://doi.org/10.1109/AINA.2018.00049
5. F. Beierle, S. Egger, MobRec—mobile platform for decentralized recommender systems. IEEE Access **8**, 185311–185329 (2020). https://doi.org/10.1109/ACCESS.2020.3029319
6. F. Beierle, T. Eichinger, Collaborating with users in proximity for decentralized mobile recommender systems, in *Proceedings of the IEEE 16th International Conference on Ubiquitous Intelligence and Computing (UIC)* (IEEE, New York, 2019), pp. 1192–1197. https://doi.org/10.1109/SmartWorld-UIC-ATC-SCALCOM-IOP-SCI.2019.00222
7. D. Camps-Mur, A. Garcia-Saavedra, P. Serrano, Device-to-Device communications with wi-fi direct: overview and experimentation. IEEE Wireless Commun. **20**(3), 96–104 (2013). https://doi.org/10.1109/MWC.2013.6549288
8. A. Crossen, J. Budzik, K.J. Hammond, Flytrap: intelligent group music recommendation, in *Proceedings of the 7th International Conference on Intelligent User Interfaces* (ACM, New York, 2002), pp. 184–185
9. N. Davies, A. Friday, P. Newman, S. Rutledge, O. Storz, Using Bluetooth device names to support interaction in smart environments, in *Proceedings of the 7th International Conference on Mobile Systems, Applications, and Services.* MobiSys '09 (ACM, New York, 2009), pp. 151–164. https://doi.org/10.1145/1555816.1555832
10. D.J. Dubois, Y. Bando, K. Watanabe, H. Holtzman, ShAir: extensible middleware for mobile peer-to-peer resource sharing, in *Proceedings of the 2013 9th Joint Meeting on Foundations of Software Engineering.* ESEC/FSE 2013 (ACM, New York, 2013), pp. 687–690. https://doi.org/10.1145/2491411.2494573
11. N. Eagle, A. Pentland, Social serendipity: mobilizing social software. IEEE Pervasive Comput. **4**(2), 28–34 (2005)
12. S. Egger. *Design and Implementation of a Mobile Platform for Recommender Systems.* Master's Thesis (Technische Universität Berlin, Berlin, 2019)
13. T. Eichinger, F. Beierle, S.U. Khan, R. Middelanis, S. Veeraraghavan, S. Tabibzadeh, Affinity: a system for latent user similarity comparison on texting data, in *Proceedings of the 2019 IEEE International Conference on Communications (ICC)* (IEEE, New York, 2019), pp. 1–7. https://doi.org/10.1109/ICC.2019.8761051

14. T. Eichinger, F. Beierle, R. Papke, L. Rebscher, H.C. Tran, M. Trzeciak, On gossip-based information dissemination in pervasive recommender systems, in *Proceedings of the 13th ACM Conference on Recommender Systems (RecSys)* (ACM, New York, 2019), pp. 442–446. https://doi.org/10.1145/3298689.3347067

15. J. Froehlich, M.Y. Chen, I.E. Smith, F. Potter, Voting with your feet: an investigative study of the relationship between place visit behavior and preference, in *UbiComp 2006: Ubiquitous Computing*, ed. by P. Dourish, A. Friday. Lecture Notes in Computer Science (Springer, Berlin, 2006), pp. 333–350. https://doi.org/10.1007/11853565_20.

16. S.M. Kala, V. Sathya, S. S. Magdum, T.V.K. Buyakar, H. Lokhandwala, B.R. Tamma, Designing infrastructure-less disaster networks by Leveraging the AllJoyn framework, in *Proceedings of the 20th International Conference on Distributed Computing and Networking*. ICDCN '19 (ACM, New York, 2019), pp. 417–420. https://doi.org/10.1145/3288599.3295596

17. K. Kwan, B. Greaves, FileLinker: simple peer-to-peer file sharing using Wi-fi direct and NFC, in *Proceedings of the 2019 IST-Africa Week Conference (IST-Africa)* (2019), pp. 1–9. https://doi.org/10.23919/ISTAFRICA.2019.8764840

18. H. Lokhandwala, S.M. Kala, B.R. Tamma, Min-O-Mee: a proximity based network application leveraging the AllJoyn framework, in *Proceedings of the 2015 International Conference on Computing and Network Communications (CoCoNet)* (2015), pp. 613–619. https://doi.org/10.1109/CoCoNet.2015.7411252

19. Z. Lu, G. Cao, T.L. Porta, Networking smartphones for disaster recovery, in *Proceedings of the 2016 IEEE International Conference on Pervasive Computing and Communications (PerCom)* (IEEE, New York, 2016), pp. 1–9. https://doi.org/10.1109/PERCOM.2016.7456503

20. J. Manweiler, R. Scudellari, L.P. Cox, SMILE: encounter-based trust for mobile social services, in *Proceedings of the 16th ACM Conference on Computer and Communications Security* (ACM, New York, 2009), pp. 246–255

21. Z. Mao, J. Ma, Y. Jiang, B. Yao, Performance evaluation of WiFi direct for data dissemination in mobile social networks, in *Proceedings of the 2017 IEEE Symposium on Computers and Communications (ISCC)* (IEEE, New York, 2017), pp. 1213–1218. https://doi.org/10.1109/ISCC.2017.8024690

22. B. McMahan, E. Moore, D. Ramage, S. Hampson, B.A. y Arcas, Communication-efficient learning of deep networks from decentralized data, in *Proceedings of the 20th International Conference on Artificial Intelligence and Statistics*, vol. 54, ed. by A. Singh, J. Zhu. Proceedings of Machine Learning Research (PMLR, New York, 2017), pp. 1273–1282

23. C. Mettouris, G.A. Papadopoulos, Ubiquitous recommender systems. Computing **96**(3), 223–257 (2014). https://doi.org/10.1007/s00607-013-0351-z

24. B.N. Miller, I. Albert, S. K. Lam, J.A. Konstan, J. Riedl, MovieLens unplugged: experiences with a recommender system on four mobile devices, in *Proceedings of the People and Computers XVII—Designing for Society*, ed. by E. O'Neill, P. Palanque, P. Johnson (Springer, Berlin, 2004), pp. 263–279. https://doi.org/10.1007/978-1-4471-3754-2_16

25. T. Oide, T. Abe, T. Suganuma, Infrastructure-less communication platform for off-the-shelf android smartphones. Sensors **18**(3), 776 (2018). https://doi.org/10.3390/s18030776

26. A.-K. Pietiläinen, E. Oliver, J. LeBrun, G. Varghese, C. Diot, MobiClique: middleware for mobile social networking, in *Proceedings of the 2Nd ACM Workshop on Online Social Networks*. WOSN '09 (ACM, New York, 2009), pp. 49–54. https://doi.org/10.1145/1592665.1592678

27. J. Rodrigues, E.R.B. Marques, L.M.B. Lopes, F. Silva, Towards a middleware for mobile edge-cloud applications, in *Proceedings of the 2Nd Workshop on Middleware for Edge Clouds & Cloudlets*. MECC '17 (ACM, New York, 2017), pp. 1:1–1:6. https://doi.org/10.1145/3152360.3152361.

28. G. Ruffo, R. Schifanella, A peer-to-peer recommender system based on spontaneous affinities. ACM Trans. Internet Technol. **9**(1), 1–34 (2009). https://doi.org/10.1145/1462159.1462163

29. A. Sapuppo, Spiderweb: a social mobile network, in *Proceedings of the 2010 European Wireless Conference (EW)* (2010), pp. 475–481. https://doi.org/10.1109/EW.2010.5483495

30. Y. Shafranovich, bluetooth data exchange between android phones without pairing (2016). arXiv:`1507.00650 [cs]`

31. J. Shu, S. Kosta, R. Zheng, P. Hui, Talk2Me: a framework for device-to-device augmented reality social network, in *Proceedings of the 2018 IEEE International Conference on Pervasive Computing and Communications (PerCom)* (IEEE, New York, 2018), pp. 1–10. https://doi.org/10.1109/PERCOM.2018.8444578

32. A. Sikora, M. Krzysztoñ, M. Marks, Application of bluetooth low energy protocol for communication in mobile networks, in *Proceedings of the 2018 International Conference on Military Communications and Information Systems (ICMCIS)* (2018), pp. 1–6. https://doi.org/10.1109/ICMCIS.2018.8398689

33. J. Teng, B. Zhang, X. Li, X. Bai, D. Xuan, E-Shadow: lubricating social interaction using mobile phones, in *Proceedings of the 2011 31st International Conference on Distributed Computing Systems* (2011), pp. 909–918. https://doi.org/10.1109/ICDCS.2011.48

34. J. Teng, B. Zhang, X. Li, X. Bai, D. Xuan, E-Shadow: lubricating social interaction using mobile phones. IEEE Trans. Comput. **63**(6), 1422–1433 (2014). https://doi.org/10.1109/TC.2012.290

35. S. Trifunovic, M. Kurant, K.A. Hummel, F. Legendre, WLAN-Opp: Ad-Hoc-Less opportunistic networking on smartphones, in *Ad Hoc Networks*. New Research Challenges in Mobile, Opportunistic and Delay-Tolerant Networks, vol. 25 (2015), pp. 346–358. https://doi.org/10.1016/j.adhoc.2014.07.011

36. O. Turkes, H. Scholten, P.J.M. Havinga, Opportunistic beacon networks: information dissemination via wireless network identifiers, in *Proceedings of the 2016 IEEE International Conference on Pervasive Computing and Communication Workshops (PerCom Workshops)* (2016), pp. 1–6. https://doi.org/10.1109/PERCOMW.2016.7457153

37. Y. Wang, L. Wei, Q. Jin, J. Ma, Alljoyn based direct proximity service development: overview and prototype, in *Proceedings of the 2014 IEEE 17th International Conference on Computational Science and Engineering* (2014), pp. 634–641. https://doi.org/10.1109/CSE.2014.138

38. Y. Wang, A.V. Vasilakos, Q. Jin, J. Ma, Survey on mobile social networking in proximity (MSNP): approaches, challenges and architecture. Wireless Netw. **20**(6), 1295–1311 (2014). https://doi.org/10.1007/s11276-013-0677-7

39. X. Wang, H. Wang, K. Li, S. Yang, T. Jiang, Serendipity of sharing: large-scale measurement and analytics for Device-to-Device (D2D) content sharing in mobile social networks, in *Proceedings of the 2017 14th Annual IEEE International Conference on Sensing, Communication, and Networking (SECON)* (2017), pp. 1–9. https://doi.org/10.1109/SAHCN.2017.7964925

40. Z. Yang, B. Zhang, J. Dai, A. Champion, D. Xuan, D. Li, E-SmallTalker: a distributed mobile system for social networking in physical proximity, in *Proceedings of the 2010 IEEE 30th International Conference on Distributed Computing Systems (ICDCS)* (IEEE, New York, 2010), pp. 468–477. https://doi.org/10.1109/ICDCS.2010.56

41. J. Yang, C. Poellabauer, P. Mitra, J. Rao, C. Neubecker, BlueNet: BLE-Based Ad-Hoc communications without predefined roles, in *Proceedings of the 2017 IEEE SmartWorld, Ubiquitous Intelligence Computing, Advanced Trusted Computed, Scalable Computing Communications, Cloud Big Data Computing, Internet of People and Smart City Innovation (SmartWorld/SCALCOM/UIC/ATC/CBDCom/IOP/SCI)* (IEEE, New York, 2017), pp. 1–8. https://doi.org/10.1109/UIC-ATC.2017.8397434

42. Z. Yu, Y. Liang, B. Xu, Y. Yang, B. Guo, Towards a smart campus with mobile social networking, in *Proceedings of the 2011 International Conference on Internet of Things and 4th International Conference on Cyber, Physical and Social Computing* (2011), pp. 162–169. https://doi.org/10.1109/iThings/CPSCom.2011.55

43. W. Zhang, H. Flores, P. Hui, Towards collaborative multi-device computing, in *Proceedings of the 2018 IEEE International Conference on Pervasive Computing and Communications Workshops (PerCom Workshops)* (IEEE, New York, 2018), pp. 22–27. https://doi.org/10.1109/PERCOMW.2018.8480262
44. C.-N. Ziegler, Semantic web recommender systems, in *Current Trends in Database Technology EDBT 2004 Workshops*, ed. by W. Lindner, M. Mesiti, C. Türker, Y. Tzitzikas, A.I. Vakali. Lecture Notes in Computer Science (Springer, Berlin, 2005), pp. 78–89

Chapter 12
GroupMusic: Recommender System for Groups

In this chapter, we present an example for a USN (ubiquitous social networking) application relating to groups of users, addressing B-Task4 (*Develop example ubiquitous social networking applications taking into account the results of the previous tasks.*). GroupMusic realizes a ubiquitous computing vision of generating music playlist for currently present users, based on unobtrusive mobile sensing of musical taste. The results presented in this chapter have been published in [3].

12.1 Motivation

The smartphone is both highly personalized and used for social interaction. With our privacy-aware social music playlist generator GroupMusic, we combine those two aspects. The tracking of individual smartphone usage can be leveraged for group scenarios. In the case of listening to music, individual taste can be recorded on the smartphone and used in group scenarios to create a group music playlist, considering the taste in music of each group member which is currently present at the same location.

To infer the musical taste of a single user, context data can help with filtering mechanisms: music listened to in a specific context might not be relevant when creating a taste model of a group. Previous work exists regarding the collection of context data in connection with data about music users listened to [1, 10]. The prototypically implemented applications are mostly based on explicit annotations by the user. The effort that is necessary for manual annotation is often criticized. Instead of such manual annotation, we follow the idea of automatically tracking the context in which what music was played back (cf. Part I).

By doing so, we are able to automatize most of the general workflow of our group music playlist generation system. Combining the possibility to automatically determine the user's context and utilizing the smartphone for social group scenarios,

F. Beierle, *Integrating Psychoinformatics with Ubiquitous Social Networking*, T-Labs Series in Telecommunication Services, https://doi.org/10.1007/978-3-030-68840-0_12

we envision *meeting*[1] scenarios—formal or informal gatherings of groups of people, with music playing, e.g., a party. In order to generate relevant playlists, users permanently and unobtrusively collect music and context data on their smartphones. When gathering for a meeting, the collected data is consolidated and preprocessed, and a recommender system can generate a music playlist based on the taste of all attending guests. Current technology already allows for a very high degree of automation in such a system. Furthermore, our system GroupMusic offers the possibility to research other areas, such as group recommendations in general or context-based recommender systems in specific.

When designing such a group music playlist generation system and collecting music playing data and context data, we are dealing with private data about the user, e.g., his/her location or smartphone usage patterns. The need for privacy awareness is increased by the fact that the data is directly used and has an immediate audible effect in the group setting. We argue that decentralized systems, which allow users to choose their service provider or set up their own host, in general have the potential to provide higher privacy awareness and help mitigate lock-in effects (cf. discussion about decentralization in Sect. 8.2 on page 84).

We designed a highly automatized system collecting data about music that was played and enriching it with context data, all without user interaction. This data is kept locally on the mobile device. The user can analyze it and decide what parts he/she wants to transmit to a server. In a meeting scenario, he/she can decide to send (parts of) his/her data to a meeting host to help generate a group music playlist for all meeting guests. This realizes the vision of highly automatized computing for group scenarios. In the following, we give some background information on context data in relation to music applications (Sect. 12.2) and detail the design and implementation of GroupMusic (Sect. 12.3). In Sect. 12.4, we evaluate our system, and in Sect. 12.5, we show an implementation of such a group music feature as an extension to TYDR.

12.2   Context Data and Music Applications

One of the ideas of our system is the idea that we can filter/preprocess a user's music listening history based on the context data associated with it. Using Dey's definition of context—anything relevant to the user or the application (cf. Sect. 3.2)—for a highly automatized music application, context is anything that might influence the user's choice of music. Using Yurur's definition, the music listened to can be viewed as a context for other applications, music listening then being, e.g., part of the *user context*. Thus, the part of our application that tracks the music listened to can effectively offer context data as a *logical sensor*. Following our own context data model in Sect. 4.2, the tracked music listening histories fall into the category of *core*

---

[1]In order to avoid overloading terms, we chose this rather formal sounding word.

functions. It is not important which app was used for listening to music but which music was played back.

Flytrap is described as a "group music environment" [5]. The music users listen to on their computers is tracked. Utilizing RFID badges, the presence of other users is detected. A group playlist is generated and the next track is decided by a voting mechanism. Since the publication of the paper in 2002, technological advances and especially the advent of smartphones allow for more complexity and automation.

SocialFusion follows the idea of collecting context data from multiple sources: online social networks, mobile phones, and nearby sensors [2]. For *SocialFusion*, one of the major challenges described is the mining of data that has been collected. The general process is collecting any available data and analyzing/mining that data for relevant information later, depending on the used application. *SocialFusion* furthermore follows the idea of individual and group recommendations based on the results that the data mining outputs. The key difference in our approach is that, instead of mining existing data that might be related, we specifically collect data in order to use it in a recommender system that creates group music playlists.

Mobile Music Genius aims to be a music player on a mobile device that automatically chooses a song according to the user's context [9]. Here, context is used to predict the taste of a single user, instead of a group of users.

Baltrunas et al. deal with the prediction of the best context for listening to a particular song [1]. For this, context data is collected explicitly with a graphical user interface. The user enters a rating for the song, activity, weather, mood, and most suitable time for listening to the song. In our application, all of that context data (with the exception of mood) is collected unobtrusively in the background and without disrupting the regular listening experience. In another paper about the collection of context data [10], activity and mood had to be annotated by the user, while some other context data was collected by the sensors of Android smartphones.

12.3 Design and Implementation of GroupMusic

In this section, we detail the requirements for GroupMusic before describing the role model. We then describe the components of our architecture and explain the three main processes.

12.3.1 Requirements

As motivated in Sect. 12.1 with the discussions about privacy awareness, a general requirement is (r1) privacy. In order to discuss privacy for our scenario, in the following, we will introduce the role model of GroupMusic. To materialize the vision of highly automated ubiquitous computing, our second general requirement is to have a high degree of (r2) automation that allows us to keep the necessary user interaction to a minimum. There is another requirement especially for the mobile component: the application should (r3) not significantly drain the battery, which is in general an important aspect for any mobile application.

12.3.2 Role Model

In Fig. 12.1, we give the role model of GroupMusic, including data flows. In the center, there is the meeting host. The meeting guest sender is the role of an attending guest that sends his/her music and context data. The meeting host uses three types of external providers. To the *music metadata service provider*, he/she sends music data in order to receive metadata about the music, e.g., genre. To the *music playlist*

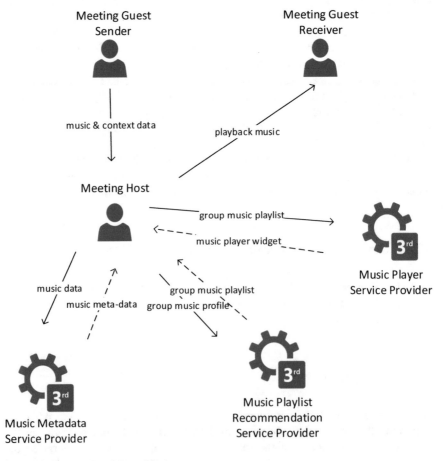

Fig. 12.1 Role model of GroupMusic

recommendation service provider, the meeting host sends a group music profile that he/she created. By utilizing a recommender service, the host ensures to create a playlist potentially containing new music previously not listened to by the attendees. To the *music player service provider*, he/she sends the (post-filtered) group music playlist he/she received from the *music playlist recommendation service provider*. The meeting guest receiver is an attending guest in the role of a listener of the played music. By hearing the music, he/she can know the generated playlist.

For the context of mobile apps, we defined privacy as the ability of the individual to choose what information someone wants to share with other users or with service providers (cf. Sect. 8.2, page 84). One part of privacy is the data a user actively shares with another user or service provider. Another aspect concerning privacy is when individual users are (unintentionally) identifiable (or traceable) through data they share. To encompass all potential aspects and data flows, our role model will help with the evaluation of the privacy of GroupMusic.

Wilson et al. describe a concept to enhance user privacy in [11]. The concept is a combination of data separation and data frugality, which we apply in GroupMusic as well: transmitting only the data that is required to use a service and using different service providers for different services bring the advantage that the multitude of external service providers cannot derive a holistic profile of a user. In the context of application development, data separation and frugality imply to build software just for a specific task with well-defined boundaries and to restrict the data collection to just the data needed for the specific task.

12.3.3   Mobile Component

Figure 12.2 depicts the components of GroupMusic. On the top, there are multiple users with their smartphones.[2] The installed meeting application contains one main component—the *Data Collection Engine*. It is responsible for the unobtrusive and continuous collection of music information of played songs. Additionally, in the background, it accesses different sensors and web services to obtain and store context data.

There are two types of collected data: music data and context data. Music data comprises the artist, track, and album. These data are generally enough to uniquely identify a song. Other music metadata like genre, tempo, etc. can be acquired later. Music playback can be tracked in two different ways. First, as we describe in Sect. 4.2, most major music apps broadcast the music that is being played back, enabling other apps to track the playback. The second option, described in Sect. 11.6.2, lets the user sign in to his/her music streaming service provider. Some of them, including Spotify, offer developers APIs to retrieve recently played back

---

[2]We developed our prototype for Android.

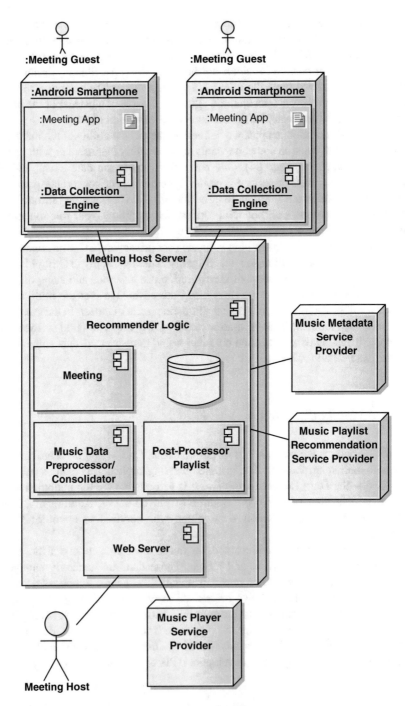

Fig. 12.2 Components of the architecture of GroupMusic

tracks. In our implementation of the group music playlist generator, we went with the first option, supporting multiple music players at once.

Applying our context data model, *device* context does not seem that useful, as we do not expect battery status or WiFi connectivity to significantly influence the choice of music. *Core functions* contain metadata from playing back music, calls, taking photos, and notifications. Again, we do not expect significant influence between music choice and the other points of metadata. Playing back music likely does not happen during calls, for example. The context category *apps* could show some relationship to what kind of music is being played back, e.g., while using the browser. The most relevant and significant context category for music playback is *physical*. Here, we expect to find correlations between location, physical activity, or weather and music playback choice. These contexts for music playback can be useful for the preprocessing step of the recommender system.

In our GroupMusic prototype, we implemented the collection of current activity[3] and weather,[4] data that related work also considered of importance with respect to the choice of music [1, 10]. For the physical activity, the most likely state returned by the SDK is stored. The possible states are sedentary, walking, running, biking, in car, random, or none. Additionally, we store the associated date and time as well as the location at which each track song was played back.[5]

12.3.4   Server Component

In the center of Fig. 12.2, we show the server component. It is used to create meetings, automatically generate group music playlists, and offer the possibility to play back music. A web server provides a website to the host. The website is the interface to the meeting host for creating meeting s and playing back music.

The *Recommender Logic* module is responsible for creating a group music playlist out of the attendees' data. We view the process of generating a group music playlist as a recommender system task. In general, for such a group recommendation task, there are two approaches [4]:

- consolidate profiles into one group profile
- consolidate individual recommendations to one group recommendation

Using an external music recommender system, the first option offers the individual guest more privacy: when the individual profile data is preprocessed and consolidated in order to generate a group music profile, only the created group music

---

[3] Via the Intel Context Sensing SDK, https://software.intel.com/en-us/context-sensing-sdk.

[4] Via OpenWeatherMap, http://openweathermap.org/.

[5] Regarding the implementation, note that we did not use the Google Awareness API, which we used for TYDR, here. Google announced that API in 2016, after the release of the paper this section is based on [3].

profile has to be given to the music playlist recommendation service provider. The individual user is thus less identifiable or traceable.

12.3.5   External Service Providers

In our prototypical GroupMusic implementation, we used three types of external service providers. As a *music metadata service provider*, returning, e.g., genre of a music track, we used Gracenote.[6] As a *music playlist recommendation service provider*, for generating a playlist from a group music profile, we used The Echo Nest.[7] The music player is integrated in our website via JavaScript; in our prototype, we used Deezer.[8]

12.3.6   Processes

This section explains the three main processes of the GroupMusic system: the meeting creation, the registration, and the meeting itself. Furthermore, the straightforward process of collecting music and context data on the mobile device is the foundation for the generation of the collaborative playlist. An extension to the current implementation could be to have an on/off switch for tracking and to offer some filtering that disables tracking for certain times, locations, contexts, or genres.

In the following, the processes are illustrated with the help of the BPMN notation.[9] Originally created for business processes, we believe BPMN is also a useful generic tool to describe software processes and interactions between users and service providers.

Figure 12.3 shows the process of creating a meeting. For the meeting creation, the meeting host installs the software on his/her own home server. In the web frontend, the host can specify date, location, and (optionally) a music genre of the meeting. This data is forwarded to the server, which in turn persists the meeting and generates a website for it. The website contains a music player widget from an external provider, which at the time of the meeting will include the continuously updated group playlist. Furthermore, the website contains a QR code for distributing to potential meeting guests.

Figure 12.4 shows the registration process. The three small, parallel lines at the bottom indicate the user as a *multi-instance participant*. This means that there are multiple users which communicate to one meeting host. The registration process

---

[6]http://www.gracenote.com/.

[7]http://the.echonest.com/.

[8]http://www.deezer.com/.

[9]https://www.omg.org/spec/BPMN/2.0.2/.

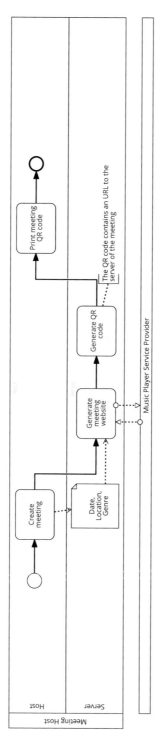

Fig. 12.3 The process of the meeting host to create a meeting

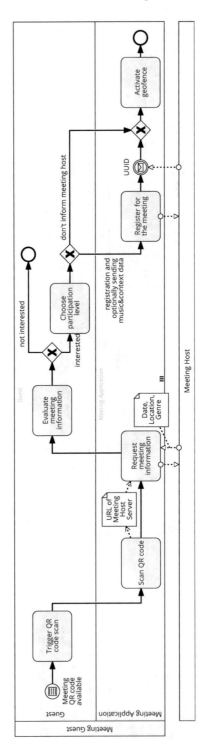

Fig. 12.4 The activities to register for a meeting

starts with the condition event that a user of the meeting application has received a QR code of a meeting. He/She triggers the scanning inside the app. The result is a URL pointing to the REST interface of the meeting host server. The application uses this URL to request some information from the server, like date, location, and genre of the meeting. The user then decides if he/she is interested in the meeting (cf. the exclusive gateway in the figure). If he/she is not interested, the process terminates. If the user is interested, he/she can decide whether to inform the meeting host about his/her interest (expressing interest is effectively an RSVP for the event). Not informing the meeting host and not sending any data would be the most private setting, but the attendee would not be enabled to participate in the creation of the group playlist. If the guest decides to attend and inform the meeting host, he/she can decide what data—if any at all—he/she later wants to send to the meeting host. The implementation so far allows for a manual song-based filtering that allows the user to remove tracks that he/she does not want to send to the host. An extension to that would be to enable the user to filter music and context data by location, context, genre, or time to remove unwanted entries more easily. When deciding to send any data to the meeting host, the guest's device receives a unique user ID (unique for the event) in order to later avoid guests sending their data multiple times.

Afterward, when being registered for a meeting, a geofence is activated in the system's location API, so the application finds out automatically if the guest is at the meeting location. Geofences are virtual boundaries for geographical areas and are used to check whether a device is entering, dwelling inside, or exiting such an area [6, 8]. The geofence is derived from the location of the meeting and a radius the host indicated when setting up the meeting. Henceforth, the system continuously checks if the user has entered the geofence. The general process of a meeting is illustrated in Fig. 12.5. The geofencing process is shown as the expanded sub-process "Meeting-entering scanning" in the top left of Fig. 12.5. If the user enters the geofence of a meeting, the application verifies that the time and date are correct. If not, the process jumps back and continues checking for the geofence. If the location and date correspond to the meeting information, the application notifies the meeting host server and sends the filtered music and context data. On the user side, the process continues with scanning, in order to check if the guest leaves the meeting. If this happens, the application notifies the server.

When the meeting host server receives attendees' data, the intermediate message event "Music&Context data" attached to the boundary of the data reception activity is called and leads to the process for creating the collaborative playlist. The process starts with the preprocessing of the data. It is depicted as a collapsed sub-process, observable by the plus sign inside the small square at the bottom center of the activity. Multiple steps are executed:

- *Augmentation* of the music data with useful metadata (e.g., genre).
- *Normalizing* the music data of one guest over all attendees: the server has to limit the influence of users with a very high amount of played tracks and a high track count in relation to a user with lower numbers.

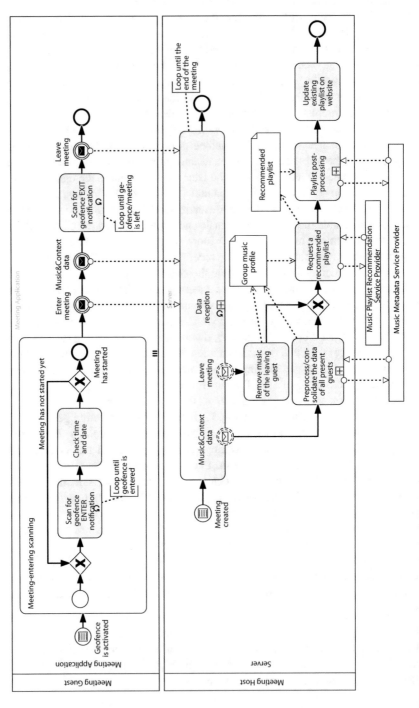

Fig. 12.5 The general process of a GroupMusic event

- *Filtering/Weighting* of the music data. Here, the context data and the music metadata is used to filter and weight music to, e.g., discard music listened to early in the morning or while exercising.
- *Consolidation* of the music data of every guest to build a *group music profile*. The result is a list that represents the preprocessed music data of all meeting attendees.

The server uses the resulting *group music profile* in the next step and hands it over to an external playlist recommendation service provider, which returns a *recommended playlist*. To avoid playing the same song or artist twice in a short amount of time, this playlist is post-processed. When this activity is finished, the website is updated, so the host is able to play the recommended songs based on the taste of the current attendees.

Looking at the "Data reception" activity in Fig. 12.5, we can see that the process for creating the recommended playlist is permanently repeated for new guests who enter the meeting. The attached "Music&Context data" message event is non-interrupting (denoted by the dashed circles around the event). It creates a new parallel process which results in the generation of the playlist. The "Data reception" process is continued until the loop ends, i.e., when the meeting is over.

When a guest leaves the meeting, the message event "Leave meeting" (attached to "Data reception") is triggered. The started process flow removes the guest's music data from the *group music profile*. Afterward, the process flow is merged with the process for generating the recommended playlist.

12.4   Evaluation

In this section, we evaluate GroupMusic according to the requirements we established.

12.4.1   Requirement 1: Privacy

Starting from our definition of privacy, in designing GroupMusic, we made sure that the user can choose what data about him/her are shared with other users and service providers. Looking at the meeting guest and the mobile application, we first examine the music and context data collection: when collecting context data, a location or weather provider will most likely not be able to distinguish between regular querying of location and weather data by other applications—especially because location and weather are highly typical data that are frequently queried. The collection of data of physical sensors and music data, assuming playing locally available music files, is completely done offline. If the user chooses to use a music streaming service provider, music preferences are of course available to that service provider. Our app still only tracks played back music locally.

Looking at Table 12.1, the "Guest Sender" role, i.e., the attending guest who sends data to the meeting host, has all music data and context data. He/She can generate filtered lists of both those datasets before sending.

We designed our system GroupMusic so that the meeting host can be anybody who wants to host a meeting. He/She installs the server component software and can start to create meetings. By this design, we generally assume a high degree of trust between meeting host and meeting guests. The meeting host receives the attendee's music and context data. He/She communicates with three external service providers. The context data remain with the host and are not sent to any service providers.

The first service provider contacted is the music metadata service provider (MMSP in Table 12.1). The filtered music data is sent to the server in order to receive additional information about the music, e.g., genre. By receiving the data, the service provider can infer something about the musical taste of the attending guests. To improve the privacy of host and guests, a local database containing music information could be introduced to remove the need to query an external service. The second service provider contacted is the music playlist recommendation service provider (MPRSP in Table 12.1). Here, the meeting host sends the locally created group music profile. By sending a group music profile, it is indistinguishable for the service provider if the request originated for one user or for a group. Individual users will most likely not be traceable in such group music profiles. The service provider can (and is supposed to by the logic of GroupMusic) infer the musical taste by this profile and returns a recommended playlist. To have more privacy for host and guests, a local recommender system could be implemented, although it might be hard to compete with services like The Echo Nest. The third service provider contacted is the music player service provider (MPSP in Table 12.1). The meeting host sends the post-processed playlist. By offering music streaming for the requested songs, the music player service provider knows the musical taste of the group. In order to increase the level of privacy for host and guests here, only locally available music files could be played.

From the perspective of the "Guest Sender" role, by looking at the checked boxes in Table 12.1, the privacy implications are indicated. The host has to be trusted with the filtered music and context data. The privacy implications of the three service providers are described above. The last step in the process of hosting a meeting is the actual playback of the music. At this step, the "Guest Receiver" role "receives" the playlist by listening to it. This is another privacy implication to be considered by the Guest Sender, as he/she might want to hide his/her musical taste. In the general case of a meeting with several attendees, it will be virtually impossible to make out a correlation between a particular song or taste in music and an individual guest. The higher the number of attendees, the more unlikely it is to make out such correlations. For very low numbers of attending guests, there are some edge cases. If very few attendees are present and a guest is arriving or leaving, the playlist should not change immediately in order to not give an indication of the individual's taste in music that could be considered private. To avoid this problem, the process of updating the

Table 12.1 Overview of roles and their data access

Role	Data					
	Music data	Filtered music data	Context data	Filtered context data	Group profile	Playlist
Guest sender	•	•	•	•		•
Guest receiver						•
Host		•		•	•	•
MMSP		•				
MPRSP					•	
MPSP						•

MMSP: Music metadata service provider
MPRSP: Music playlist recommendation service provider
MPSP: Music player service provider

playlist can be done in bulk after several people entered/left, in order to hide an individual's taste in a group of people. If there are zero (or one, if the host is also a guest) attendees, the genre setting of the meeting can still make it possible to play back music.

Overall, the employed mechanisms of filtering data on the smartphone before sending, sending limited data to service providers, and considering edge cases with few attendees leave the user in control of his/her data and avoid unintentional sharing of private data. In future work, the proper way of presenting the user the information given in Table 12.1 could be researched. To lower the level of trust needed toward the meeting host, some parts of the preprocessing of the data could be done on the smartphone. This way, the meeting guest's privacy could be improved because he/she does not send the list of music tracks listened to with associated context data but some form of music profile that models his/her taste.

12.4.2 Requirement 2: Automation

Looking at the mobile component, the collection of music and context data is done without any user interaction. The user just uses his/her music player application. Music preferences are expressed by the listening behavior.

The meeting host just has to create a meeting and can use the automatically created QR code to invite meeting guests. The "Host" part of Fig. 12.3 shows this. At the meeting itself, he/she just needs to start the music player widget on the automatically generated meeting website.

The top part of Fig. 12.4 visualizes all these manual interactions necessary for registering for an event. The meeting guest has to trigger the QR code scan. After receiving the meeting information from the server, he/she can decide whether he/she wants to attend and create a filtered list of music and context data to send. Choosing what data to send is a necessary manual step in order to assure the guest's privacy. Without considering privacy, this step could be automated, and every user would send their data.

The meeting itself can then run fully automatically. Coming guests are automatically considered, and leaving guests are automatically disregarded in the group music playlist. Note that in Fig. 12.5, neither host nor guest are shown, indicating the fully automatized process.

12.4.3 Requirement 3: Battery Consumption

The mobile application is used in two processes: tracking music and context and sending data to the meeting host. The latter process is executed only occasionally. The tracking of music and context data is run permanently when listening to music, so we focus on this process in this evaluation. We used an LG Nexus 5, running stock Android 5.1.1 with only the stock Google applications and an additional file explorer installed. We collected two sets of data. In both cases, during the whole run, we turned off the display and let music play continuously. The first dataset tracked the battery life without using our application and yielded 19.5 h. For the second dataset, we let our mobile application run. The battery lasted 19.3 h. Our evaluation shows that the usage of our application made the battery drain only around 1% faster. Our interpretation of this result is that the power consumption by playing back music is already so high that the tracking of music and context data does not carry a lot of additional weight. Furthermore, in real-world scenarios, additional applications that cause traffic and notifications, like email, messaging, or online social network applications, will most likely make the additional battery drain by our application unnoticeable.

12.5 Group Playlist Generation as a TYDR Extension

In the course of his bachelor's thesis, Jan Pokorski implemented a proof-of-concept prototype showing the same idea without a server component, running fully on the smartphone [7]. For this, we developed an extension for TYDR called TYDR Fusion. This extension contains new tiles that contain the host and guest components for group recommendations for music and POIs (points of interest).

Figure 12.6 shows screenshots of TYDR Fusion. Figure 12.6a shows the extended main screen. Extending the "Group Music" tile reveals two buttons "Host" and "Share." The host presses the "Host" button and sees a waiting screen with the number of people who are connecting. The guest who chooses the "Share" option can select the music tracks he/she wants to share. Once the host confirms that all attendees joined, a group playlist is generated and sent to all participants (see Fig. 12.6b). Similar to the work described above, the first step of the recommenda-

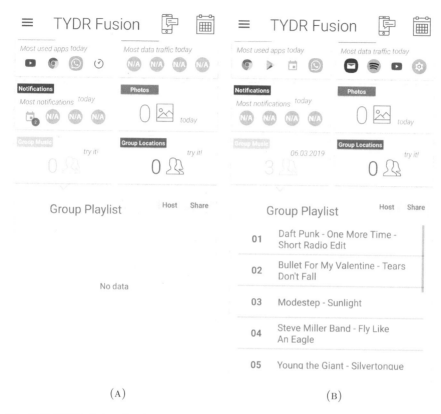

Fig. 12.6 TYDR Fusion showing host and guest module running on the smartphone. (**a**) TYDR extension main screen. (**b**) Resulting music recommendations

tion process is preprocessing of the data. The goal is a group profile, in this case a list of seeds that can be entered in the Spotify API to retrieve recommendations (the same music recommender endpoint we use in our MVP (minimum viable product) for the mobile platform for decentralized recommender systems MobRec (cf. Sect. 11.6.4)). The titles in the resulting group playlist can be tapped, which opens a YouTube search for the title. For device-to-device communication, we used Google Nearby Connections. Additionally to music group recommendation, we also implemented POI recommendations, using Foursquare.

12.6   Summary

In this chapter, we combined the ideas of the smartphone being both a highly individualized device and a device for social interaction. We designed and implemented a group music playlist generator GroupMusic that continuously creates group music playlists based on group music profiles for guests attending a meeting. We use the smartphone to track individual musical tastes. We evaluated the privacy, automation, and battery consumption of GroupMusic. When designing GroupMusic, we took into consideration privacy awareness and lock-in effects. We keep the data collected to a minimum with respect to the application it is used for. We also detailed further possible improvements to advance the level of privacy even more. The level of automation is already very high, and further automation might only be possible by compromising privacy. For the mobile application, we evaluated that the additional battery consumption is negligible. The design of GroupMusic might give hints on how to develop other privacy-aware USN applications or group recommender systems.

Future work includes evaluating the preprocessing and consolidation step for generating the group music profile to quantify the quality of the generated playlists. Furthermore, the collected data could be used for other features like displaying who is attending the party or for visualizing a map which displays where which music is popular. Current technologies also allow for even more automation: BLE (Bluetooth Low Energy) beacons could be used for accurate indoor positioning of meeting guests in order to automatically infer the atmosphere of the party. The smartphone accelerometer could be utilized for recognizing dancing moves to additionally impact the social music playlist creation.

References

1. L. Baltrunas, L. Rokach, M. Kaminskas, B. Shapira, F. Ricci, K.-h. Luke, Best Usage context prediction for music tracks, in *In Proceedings of the 2nd Workshop on Context Aware Recommender Systems* (2010)
2. A. Beach, M. Gartrell, X. Xing, R. Han, Q. Lv, S. Mishra, K. Seada, Fusing mobile, sensor, and social data to fully enable context-aware computing, in *Proceedings of the Eleventh Workshop on Mobile Computing Systems & Applications* (ACM, New York, 2010), pp. 60–65
3. F. Beierle, K. Grunert, S. Göndör, A. Küpper, Privacy-aware social music playlist generation, in *Proceedings of 2016 IEEE International Conference on Communications (ICC)* (IEEE, 2016), pp. 5650–5656. https://doi.org/10.1109/ICC.2016.7511602
4. S. Berkovsky, J. Freyne, Group-based recipe recommendations: analysis of data aggregation strategies, in *Proceedings of the Fourth ACM Conference on Recommender Systems, RecSys '10* (ACM, New York, 2010), pp. 111–118. https://doi.org/10.1145/1864708.1864732
5. A. Crossen, J. Budzik, K.J. Hammond, Flytrap: intelligent group music recommendation, in *Proceedings of the 7th International Conference on Intelligent User Interfaces* (ACM, New York, 2002), pp. 184–185
6. A. Küpper, U. Bareth, B. Freese, Geofencing and background tracking—the next features in LBSs, in *Informatik 2011—Informatik Schafft Communities*. LNCI, vol. 192 (2011), p. 14
7. J.S. Pokorski, TYDR fusion: ubiquitous social networking based on automatically tracked data. Bachelor's Thesis. Technische Universität Berlin (2019)
8. S. Rodriguez Garzon, B. Deva, Geofencing 2.0: taking location-based notifications to the next level, in *Proceedings of the 2014 ACM International Joint Conference on Pervasive and Ubiquitous Computing. UbiComp '14* (ACM, New York, 2014), pp. 921–932. https://doi.org/10.1145/2632048.2636093

9. M. Schedl, G. Breitschopf, B. Ionescu, Mobile music genius: reggae at the beach, metal on a Friday night?, in *Proceedings of International Conference on Multimedia Retrieval* (ACM, New York, 2014), pp. 507–510
10. Y.-C. Teng, Y.-S. Kuo, Y.-H. Yang, A large in-situ dataset for context-aware music recommendation on smartphones, in *2013 IEEE International Conference on Multimedia and Expo Workshops (ICMEW), July* (2013), pp. 1–4. https://doi.org/10.1109/ICMEW.2013.6618254
11. C. Wilson, T. Steinbauer, G. Wang, A. Sala, H. Zheng, B.Y. Zhao, Privacy availability and economics in the Polaris mobile social network, in *Proceedings of the 12th Workshop on Mobile Computing Systems and Applications. HotMobile '11* (ACM, New York, 2011), pp. 42–47. https://doi.org/10.1145/2184489.2184499

Chapter 13
Interim Conclusions for Part II: Advancing Ubiquitous Social Networking Through Mobile Sensing

Part II addressed research question B (*How to use insights from psychological research for the design of social networking applications?*). Based on our research in Part I, we utilized the technical approach of mobile sensing and the conceptual finding of associations between smartphone data and psychological structures. Psychological research and research in social sciences show us that *homophily*, the tendency of forming of bonds between similar people, and *propinquity*, the tendency of forming bonds between people in proximity, are crucial factors for the formation of connections in real-life social networks. Not only does smartphone data reflect personality but also preferences, for example, relating to locations or music. This property makes data from mobile sensing especially useful for finding similarities between users, as well as using such data for recommender systems. In Part II, we researched and developed concepts, metrics, and services for unobtrusive USN (ubiquitous social networking) scenarios based on device-to-device communication. For users of systems based on our results, the main benefits are potentially more meaningful recommendations, unobtrusive mobile sensing without manual user input, and fewer lock-in effects compared to existing recommender systems. For researchers and software developers, our concepts, metrics, and prototypical implementations serve as blueprints for implementing unobtrusive systems with seamless multiplatform (Android/iOS) device-to-device communication on off-the-shelf devices.

For the incentivization of social interaction between strangers, we developed our approach SimCon, which is based on the abovementioned principles of homophily and propinquity. According to SimCon, similarity in smartphone data collected via mobile sensing is the deciding factor for contact recommendations. In order to closely estimate the similarity between two users, we introduced metrics for similarity estimation, most importantly CBF-Dice for the comparison of two Counting Bloom filters. We utilize this similarity estimation for finding similar users in our platform for decentralized recommender systems MobRec. The similarity estimation and, with it, the SimCon concept are implemented as extensions to

F. Beierle, *Integrating Psychoinformatics with Ubiquitous Social Networking*,
T-Labs Series in Telecommunication Services,
https://doi.org/10.1007/978-3-030-68840-0_13

TYDR (Sect. 10.7), showing that mobile sensing is a useful technique not only for psychoinformatics but also for advancing USN. This is true for services offered for individual users (MobRec) or pairs of user (SimCon) and also for group scenarios: our group recommender system GroupMusic is based on mobile sensing as well, and we prototypically implemented it as an extension to TYDR as well (Sect. 12.5).

Part III
Conclusions and Outlook

Chapter 14
Conclusions

The ultimate goal of this thesis is related to advancing and improving the understanding and well-being of people. In order to deepen the scientific understanding of people, we developed the psychoinformatical mobile crowdsensing app TYDR (Tack Your Daily Routine). TYDR tracks a large variety of sensor data sources and smartphone usage statistics. A standardized psychometric questionnaire assesses the user's personality traits. Future psychoinformatical research with TYDR is already planned with research partners in Germany and Austria, tracking smartphone usage of large samples of patients. To the best of our knowledge, with our privacy model PM-MoDaC, we are the first to propose and implement a full-scale privacy model for mobile data collection apps. PM-MoDaC consists of nine concrete measures that researchers and developers can implement to protect the privacy of their users. PM-MoDaC already gained traction in the research community and serves as a basis for privacy awareness in related studies.

Overall, through our international and interdisciplinary research collaborations, we could attract almost 4000 TYDR users that contributed smartphone data. We evaluated our privacy model and analyzed what data users are willing to share with researchers. For instance, as one of our results, we found that younger and female users tend to be less willing to share data. Furthermore, the type of app might influence the type of users it will attract, in terms of personality traits. These results are helpful for other researchers, especially for the study planning phase of future research in the fields of mobile crowdsensing and psychoinformatics. The knowledge we created by analyzing the data we collected with TYDR reveals the relationship between personality, demographics, and smartphone usage frequency and session duration. To the best of our knowledge, we are the first to look into these variables specifically. In our sample ($n = 526$), the users unlocked their smartphone 72 times per day on average, with a mean usage session duration of 3.7 min. We found that younger users used their smartphones more frequently but with shorter session durations. Female users have longer sessions than males users on average. The personality traits extraversion and neuroticism are associated

© The Author(s), under exclusive license to Springer Nature Switzerland AG 2021   189
F. Beierle, *Integrating Psychoinformatics with Ubiquitous Social Networking*,
T-Labs Series in Telecommunication Services,
https://doi.org/10.1007/978-3-030-68840-0_14

with more frequent smartphone use, while conscientious users have shorter session durations. These results are crucial for psychologists and mHealth (mobile health) app developers, for example, when dealing with smartphone overuse. Researchers and software developers in the fields of context-aware computing can utilize our findings for understanding their user base better or for tailoring services to specific personality types.

We applied the concepts of mobile sensing and the insights about the relationship between personality and smartphone usage to ubiquitous social networking. With SimCon, we developed a concept for the fully automated incentivization of social interaction with other users in proximity. At the core of SimCon is the finding that similarity is of key importance for the formation of interpersonal relationships. The idea of SimCon is that similar personality traits and similar preferences are reflected in similarities of tracked smartphone data. New similar contacts in proximity can be recommended fully automatically, without manual user input and without the use of any social graph that is typically utilized in new contact recommendations in social networking scenarios. Employing our concept will potentially yield more meaningful contact recommendations and lower the inhibition threshold of using a smartphone-mediated social interaction incentivization system by being fully automated. In order to closely estimate the similarity between two users, we introduced metrics for the comparisons of two multisets based on the probabilistic data structure Counting Bloom filter. A single exchange of a small amount of data suffices to accurately approximate the similarity of two users, which we evaluated on real music listening histories. Our approach is generalizable, and our comparison metric CBF-Dice can be applied for any other similarity estimation of frequencies represented as multisets.

For the development of ubiquitous social networking prototypes, we argued that decentralized systems allow for more user control about sensitive data and allow for the mitigation of lock-in effects. The user is given the choice of what data to share with which service providers. Following this concept of privacy awareness, we developed two applications for ubiquitous social networking scenarios. The first one, MobRec, is a platform for decentralized mobile recommender systems. The second one, GroupMusic, is a ubiquitous system that plays back music for groups of users. MobRec uses mobile sensing for tracking user preferences. Preference data is exchanged in a device-to-device manner which we implemented for seamless background transfer for off-the-shelf Android and iOS smartphones. Recommendations for a variety of domains, e.g., locations or music, are then calculated locally or by contacting third-party service providers. GroupMusic realizes a ubiquitous computing vision. The highly automated system plays back music according to the taste of currently present users and automatically adapts playlists when the present group changes. The system is based on mobile sensing and privacy-aware sharing of data with a host. Researchers and software developers can use our designs as blueprints for the development of seamless ubiquitous systems in the fields of device-to-device computing and social networking. Users of future systems based on our blueprints benefit from unobtrusiveness, increased privacy awareness, and freedom from lock-in effects.

Chapter 15
Outlook

Relating to psychoinformatical apps like TYDR, we see three main directions for future work: (a) technical, (b) mHealth apps, and (c) data analytics. On a (a) technical level, wearable devices could be integrated into the mobile sensing system. They could yield more accurate data about steps, monitor the user's heart rate, etc. On a software level, the development of apps like TYDR needs to take into account the changes in the underlying mobile operating systems. For example, in February 2020, Google announced that location tracking will be more restrictive in future Android version 11.[1]

Following the idea of (b) mHealth, TYDR could be extended to give users more explicit feedback about their behavior. Tracking sensor data combined with continual user feedback, the app could learn that certain actions or locations are beneficial for the user and could recommend such activities or locations. In the current TYDR implementation, there is a daily *personality states* questionnaire. It can be filled out between 6 pm and 2 am every day. Extending this concept more toward the idea of ecological momentary assessment (EMA), querying the user depending on specific context conditions, could enable the development of new mHealth apps and studies.

Regarding (c) data analytics, there are several further possibilities of research to be conducted with data collected with TYDR. We are currently working on a deep-learning-based system that predicts the next app that is going to be used, which can help in memory management and preloading of apps. Additionally, we are currently working on multiple topics related to the scientific understanding of users in terms of behavior and psychological traits. We are analyzing the relationship between personality traits and music listened to and the relationship between the user's *resilience* and his/her use of communication-related apps, and we are predicting personality traits based on TYDR data.

---

[1] https://android-developers.googleblog.com/2020/02/safer-location-access.html.

© The Author(s), under exclusive license to Springer Nature Switzerland AG 2021
F. Beierle, *Integrating Psychoinformatics with Ubiquitous Social Networking*,
T-Labs Series in Telecommunication Services,
https://doi.org/10.1007/978-3-030-68840-0_15

A point of future work relating to ubiquitous social networking is the integration of all developed concepts and prototypes. Imagine a group of users coming together and our system GroupMusic playing back music according to the taste of the currently present users. Once the users are at the same locations, we can estimate their similarity utilizing CBF-Dice and give contact recommendations according to SimCon. Furthermore, by exchanging data with the other present users, local databases can collect more data for the decentralized mobile recommender systems MobRec. While we individually evaluated each prototype, the integration of all systems and real-world deployment remains for future work.

At the time of writing, another aspect of ongoing and future work highlights another overlap between the two parts of this thesis. During the ongoing SARS-CoV-2 coronavirus pandemic that started at the end of 2019, there is an ongoing discussion of utilizing smartphones for tracing the potential spread of the virus. The virus spreads in human-to-human contact. Google and Apple joined efforts in developing a contact tracing mechanism utilizing Bluetooth.[2] Such contact tracing is relevant for the given virus pandemic, and at the same time, such Bluetooth proximity recognition is the exact moment where data transfer via Bluetooth is desired in our social networking prototypes. The technical innovations coming out of this could be used for other mHealth scenarios as well as for ubiquitous social networking scenarios.

---

[2]https://www.apple.com/newsroom/2020/04/apple-and-google-partner-on-covid-19-contact-tracing-technology/ and https://www.blog.google/inside-google/company-announcements/apple-and-google-partner-covid-19-contact-tracing-technology/.

Index

© The Author(s), under exclusive license to Springer Nature Switzerland AG 2021
F. Beierle, *Integrating Psychoinformatics with Ubiquitous Social Networking*,
T-Labs Series in Telecommunication Services,
https://doi.org/10.1007/978-3-030-68840-0

Printed in the United States
by Baker & Taylor Publisher Services